San Antonio

SAN ANTONIO

A Tricentennial History

CHAR MILLER

Texas State Historical Association
Austin

Number 25 in the Fred Rider Cotten Popular History Series.
Cover image courtesy of the San Antonio Conservation Society.

Names: Miller, Char, 1951– author.
Title: San Antonio : a tricentennial history / Char Miller.
Other titles: Fred Rider Cotten popular history series ; no. 25.
Description: Austin : Texas State Historical Association, [2018] | Series: Number 25 in the Fred Rider Cotten popular history series
Identifiers: LCCN 2018016217 | ISBN 9781625110497 (pbk.)
Subjects: LCSH: San Antonio (Tex.)--History. | San Antonio (Tex.)—Social conditions. | San Antonio (Tex.)—Environmental conditions.
Classification: LCC F394.S21157 M55 2018 | DDC 976.4/351—dc23
LC record available at https://lccn.loc.gov/2018016217

For Rebecca, Ben, & Caitlin

Contents

Acknowledgments

No historian works alone; we depend heavily on the kindness of colleagues, friends, family, and strangers. So, too, with this book, which would not be in your hands without the aid of so many others. My greatest debt is to Ryan Schumacher at the Texas State Historical Association, who approached me about the possibility of writing this volume and then proved a most generous, shrewd, and collaborative editor. Related thanks go to the Association's chief historians, Mike Campbell, and his successor, Walter Buenger, for their steadfast support of this project, and to the two anonymous readers who reviewed the manuscript prior to publication. Thanks are also due to TSHA Assistant Editor Nicholas K. Roland for his help in proofreading and editing. Anyone who writes about San Antonio's profound connection to the Edwards Aquifer will understand why I am so grateful to Gregg Eckhardt. He has poured his heart and soul into gathering, curating, and digitizing seemingly everything related to the aquifer, and then posting it on edwardsaquifer.net, an indispensable resource; Gregg has generously provided many of the images that illustrate this book. I am grateful as well to the other institutions that provided illustrative material, including the John Carter Brown Library at Brown University; the Portal to Texas History and the University of North Texas Libraries; State Preservation Board, Austin; Bexar County Commissioners Court; the Witte Museum; University of Texas at San Antonio Special Collections; San Antonio Conservation Society; the Dolph Briscoe Center for American History at the University of Texas at Austin; Tarrant County College District Archives; the University of Texas Libraries; and Texas Lutheran University. Thanks also to Daniel Galindo for the map of San Antonio's trolley lines. Deep thanks go to my home institution, Pomona College, for the gift of year-long sabbatical that enabled me to write this tricentennial history, and for a Faculty Research Grant that facilitated its production. Many colleagues at the Claremont Colleges Library contributed their time and energy to this project, most notably the hardworking staff in the interlibrary loan office. As always, Judi Lipsett let me pile books up everywhere in our Claremont home, as she once did when we lived in Olmos Park. I am thrilled to dedicate this history of San Antonio to Rebecca, Ben, and Caitlin, who grew up in the Alamo City and have never outgrown their love of the Spurs.

Prologue

To recover the past is to know the present. That claim seems particularly apt at those moments when a city celebrates a milestone occasion, much as San Antonio did in 2018 when it marked its 300th anniversary. The dating of the community's founding, which establishes the three-century-long arc of its settled existence, is only one of the topics that this brief historical survey takes into account. Certainly, the Payaya and their hunting and foraging ancestors, who long pre-dated Spanish settlement of the San Antonio River valley, might take exception to a celebration timed solely to European colonization of South Texas. But the creation of a colonial outpost, subordinating the indigenous people, and integrating them into the rigors of a sedentary life framed around religious liturgy and bounded within mission walls, is essential to the city's history.

Also essential is the equally fraught process whereby the Spanish themselves struggled to maintain their hold in their empire's northern frontier. With varying degrees of success, they fended off the Apache and Comanche, the French and Americans, who at different times pressed in on their territory across the early eighteenth and nineteenth centuries. When an internal revolt finally toppled Spain's North American empire, the founding of the Mexican Republic did not bring regional stability, either: San Antonio, as had been true throughout its first century, was at the epicenter of the battles that raged during the early decades of its second century. The Mexican revolutionary conflicts in the 1820s, the legendary 1836 siege at the Alamo, and the continuing strife with the Comanche nation only set the stage for the city's significant contributions to the U.S.-Mexican War (1846–48) and to the Civil War (1861–1865). More significant still was its support of those global conflicts that followed, from the

Spanish-American War (1898) to World War I, from World War II to the Cold War and beyond. Peace has been a rare commodity in South Texas; counterintuitively, conflict has had its benefits. One of them is that war and defense spending has helped power San Antonio's growth and development for more than 150 years.

Social tensions, partisan disputes, and violent clashes have also marked the city's history. Colonial politics were as divisive as were those that followed the rise of the Anglo power elite that came to dominate the community in the wake of the Texas Revolution, the absorption of the Texas into the United States, and ever since in the American city. Tejanos, a new social type forged in the crucible of the Spanish and Mexican eras, remained a critical presence in San Antonio after annexation, but they became second-class citizens in the face of American hegemony. Even as whites took control of the city's politics, economic resources, and social status, the process was never unilateral, complete, or uncontested; however subordinated, Tejanos resisted. With the arrival of new forms of transportation—railroads and streetcars, and later automobiles—new set of conflicts emerged. Each national crisis, or spike in population, has brought with it a slew of challenges that have shaped the local quality of life and complicated the capacity of city hall, or any entity, to meet these challenges. From the nineteenth century to the twenty-first, perhaps the most consistently difficult issue has revolved around water. The Spanish selected the location of San Antonio because of its bountiful presence. Ever since, whether because of devastating floods (and their belated control), drought, or declining quality, the community has battled over its fatal or life-affirming flow, over the ownership, control, and distribution of water.

These long-running disputes—racial conflict, ethnic strife, educational inequities, political oppression, environmental damage, and spatial segregation—are among an interlocking set of social ills that in different ways have plagued San Antonio from the beginning. Knowing this fraught, fascinating, and frustrating past may help those calling San Antonio home during its fourth century to build a more just and equitable community. It is my hope too that this short book will encourage someone to write a full and comprehensive urban biography of San Antonio—a long overdue project.

1

YANAGUANA

Yanaguana Garden, a 4.1-acre playground nestled within Hemisfair Park in downtown San Antonio, and which was dedicated with considerable éclat in October 2015, is a fascinating site for several reasons. Years earlier, it had been part of a larger area selected for HemisFair '68, a late-20th-century initiative designed to celebrate the city's 250th anniversary. Using federal monies to bulldoze some of San Antonio's oldest neighborhoods, the fair's promoters erected a world's-fair-like assemblage of exhibit halls, theaters, and arenas that they hoped would bring millions of new visitors to San Antonio, boosting the local economy and the city's visibility as a major American city. At that celebratory moment in 1968, the present took precedence over the past, and the same was true years later when Yanaguana Garden was carved out of HemisFair's ghostly remains. The playground makes no mention of the Payaya people from whose language its appellation derives and who gave the South Texas metropolis its first recorded name: Yanaguana.

The Payaya's absence from this particular recreational landscape, which is consistent with their essential erasure from the region's historical narrative, is also not particularly surprising. Their physical presence, along with the indigenous landscape they created in and around San Pedro Creek and the San Antonio River, did not survive the late eighteenth century. This was the partial result of the complex clash between the Spanish Empire as it pushed north out of Mexico and into present-day Texas and the Lipan Apache, who were moving south into the Edwards Plateau in reaction to the expansionist Comanche nation that had occupied their ancestral territory to the north. The Payaya, the area's principle indigenous group, fell prey to these oft-violent shifts in regional geopolitics.[1]

One herald of the Payaya's troubled future occurred on June 13, 1691, which appears to have been the first time that their existence was recorded in the Spanish language. That day, an entrada (exploratory expedition) commanded by Domingo Terán de los Ríos, with Fray Damian Massanet as its spiritual leader, reached a Payaya encampment in the future San Antonio. The Spanish had marched in that day from the Medina River, roughly twenty miles to the south, across what Terán described as "a fine country with broad plains—the most beautiful in New Spain." In search of trees, vertical markers of water and people, they halted close to San Pedro Springs, headwaters of the eponymous creek. As interested as Terán was in the site's natural resources—he did not miss the value of the ample flow of its clear waters or the large number of oak, cedar, willow and cypress growing in the fertile soil—he was especially struck by the human community that clustered around the bountiful springs. "Here we found certain *rancherías* in which the Peyaye [Payaya] nation live," he wrote in his diary. "We observed their actions, and I discovered that they were docile and affectionate, were naturally friendly, and were decidedly agreeable towards us." That impression led him to anticipate how the Payaya could be absorbed into and serve the Spanish imperium: "I saw the possibility of using them to form *reducciones* [evangelized villages/missions]—the first on the Rio Grande, at the presidio, and another at this point. Different nations in between could be thereby influenced."[2]

Fray Massanet was of the same mind. For him, the first step was to convert the space: "In the language of the Indians, it is called Yanaguana," he wrote. Yet even as the Franciscan missionary transcribed the Payayan term for this well-watered terrain, he effectively wrote the word out of history: "I called this place San Antonio de Pádua, because it was his day," an act of colonial conquest repeated wherever Europeans landed in the New World.

Massanet's second step was related to his first: in hopes of winning the hearts and minds of the Payaya, that is, to convert them to Catholicism, the next day, he and Terán orchestrated an impressive display of Spanish religious ritual and military might:

> I ordered a large cross set up, and in front of it built an arbor of cottonwood trees, where the altar was placed. All the priests said mass.

> High mass was attended by Governor Don Domingo Terán de los Ríos, Captain Don Francisco Martínez, and the rest of the soldiers, all of whom fired a great many salutes. When the host was elevated, a salute was fired by all the guns. The Indians were present during these ceremonies. After Mass the Indians were given to understand through [an interpreter] that the Mass and salutes fired by the Spaniards were all for the honor, worship, and adoration we owed to God, our Lord, in acknowledgment of the benefits and great blessing that His Divine Majesty bestows upon us; that it was to Him that we had just offered sacrifices in the form of the bread and wine which had just been elevated in the Mass.[3]

As the smoke cleared at ceremony's end, Massanet distributed gifts sacred and secular—"rosaries, pocket knives, cutlery, beads, and tobacco"—that underscored the benefits that could come from a close association with the missionaries. This was an exchange and accommodation that the Payaya leadership apparently understood. Its "captain," to whom Massanet also presented a horse, in turn promised to guide the expedition on its northeasterly trek and to supply the Spanish with food and labor. Terán may have believed this generosity was a mark of the Payaya's deference to Spain's power and piety, but such cultural interactions are never one-sided. The Payaya's engagement with this entrada—and those that followed—was also a choice, a reflection of their deliberate attempt to protect themselves from other, more powerful native people then dominating the region.

The Payayan perspective was rooted in their longstanding and good life fashioned within the rumpled hills, wide river valleys, and sweeping grasslands of south-central Texas. They took full advantage of variations in regional topography, its uplands and low, and made use of its diverse ecosystems, whether riparian or terrestrial, woodland or prairie. Foraging cultures had utilized plant and animal life in the area for more than 10,000 years before the Spanish had arrived, and the Payaya would be the last of those to richly exploit local resources: mammals of all sizes, snakes and other reptiles, roots, seeds, and grasses, nuts—especially pecans—and fruit. "We ought to understand," archaeologist Karen Stothert has observed about the efficiency of the hunting and gathering practices of the Payaya and their predecessors, that they "were able to support themselves *because* they defined such a wide range of biomatter as food."[4]

The Payayas' self-sufficiency depended on their mobility. To reap the benefits of seasonal plenty, they and other tribal bands moved across the landscape, following migratory animals such as bison, harvesting prickly pears when they flowered, and gleaning mescal beans, blackberries and mulberries as they ripened. In seasons when food was scarce, they nursed their hunger. So recalled Álvar Nuñez Cabeza de Vaca, who in 1528 was shipwrecked on the Texas coast, and for the next eight years lived with and traded between the Karankawas, living around Galveston Bay, and other Coahuiltecan-speaking people such as the Payaya across southern Texas. Because "nothing is planted for support," he observed, there were many days during which his hosts went without a meal, but they did not cavil or starve. Rather, to "revive our spirits, [they] would tell us not to be sad, that soon there would be prickly pears, when we should eat a plenty, and drink of the juice, when our bellies would be very big, and we should be content and joyful."[5] Mindfulness could tamp down want.[6]

Intentionality was also a strategic component of the Payayas' peregrinations: they made certain that each of their sites, which they occupied routinely if not quite permanently across a year, was close to water. It was not by happenstance that Terán and Massenet encountered the Payaya at San Pedro Springs. For millennia, native peoples had clustered near this headwater, as they had others strung along what geographer James F. Peterson has called the Texas Spring Line. Running from present-day Del Rio to the southwest of San Antonio and curving north and east to Austin, these springs, bubbling up from the Edwards Aquifer, were environmental lifelines for all species in the region.[7]

As with other foraging cultures, the Payaya carefully managed their seasonal exploitation of terrestrial and riverine resources. One aspect of their conservation strategy was evident in their calculated crisscrossing of territory rich in plants and animals, which lessened their overall impact across time and space. Fire was another tool for survival. As with many of the native people that the Spanish would encounter from Florida to California, the Payaya used fire to insure the maintenance of preferred flora and fauna. Cabeza de Vaca was an uneasy witness to this early form of prescribed burning. "The Indians of the interior," he wrote, "go with brands in the hand firing the plains and the forest within their reach . . . to drive out lizards and other like things from the earth for them to eat." He thought their indiscriminate use of fire "intolerable," but missed the beneficial, landscape-scale consequences. Not only did the flames drive game toward hunters at a place of their choosing but these annual burns produced nitrogen-rich soil and a set of fire-adapted habitats, with an array of select species that inhabited them, creating indigenous landscapes that sustained aboriginal life. Centuries later, Vinton Lee James drew on the memories of settlers from the Spanish, Mexican, and U.S. eras to conclude that South Texas Indians had ignited "immense" fires each autumn. Once touched off, they "would sweep the entire country until stopped by a river or some natural barrier, leaving an appearance of utter desolation." All was not destroyed, however, because the infernos created "a treeless prairie, covered with a rank growth of grass" that provided a "pasture ground for immense herds of deer, graceful antelope, wild horses, and the uncouth and ponderous buffalo."[8]

Spanish missionaries observed similar behavior in Alta California. The Chumash people of the central coast, like the Miwok of the Central Sierras, routinely put flame to ground, a charring that kept open country open and privileged those plants and animals that could live within, take advantage of, and in some cases required fire to regenerate. So effective was this pattern of fire management that later explorers, visitors, and naturalists gushed over California's divine, Garden of Eden-like character, little suspecting its human origin.[9] So it was in South Texas. When the well-traveled and keen-eyed Cabeza de Vaca opined that the native people who lived inland close to the spring line were more productive than their coastal peers, and were in "better condition," he was acknowledging the crucial role that human agency played in setting the parameters and possibilities of their life in the region.[10] A century and a half later, when Domingo Terán de los Ríos rode across what he described as South Texas's lush prairie, he was also unknowingly complimenting the environmental management practices of the Payaya and the other Coahuiltecan people there. They made the terrain that shaped them.

Despite their beneficial management of the land, the Payaya and their peers were unable to fully deflect or control the threats other people posed to their way of life. In a sense, the Spanish were the least of their worries, maybe even a complicated, if partial, answer to some of them. The most pressing of the Payaya's concerns were the horseback warriors of the Lipan Apache and the Comanches—a reasonable concern that came with an ironic twist: the Spanish had reintroduced the horse to North America in the sixteenth century, and by the eighteenth it had revolutionized Native life on the Great Plains. "The region that the Spaniards called Texas...differed from many of New Spain's frontiers," observes historian Julianna Barr, "because Indians not only retained control over much of the province but also because they asserted control over the Spanish themselves. The region's eighteenth-century history is not one of Indian resistance, but of Indian dominance."[11] The Caddos dominated the eastern portion of Texas; the Comanches and Wichita controlled the north; and the west was Apache territory. Only the south was (or would become) Spanish territory, and even there the Spaniards' control was in dispute. Barr notes that "for much of the eighteenth century Spaniards earned native attention only as targets of raids," a targeting that revealed their inferior position relative to those arrayed against them. As they sought a toehold in southern Texas, Spanish civil, religious, and military authorities immediately "encountered already existent and newly emerging native systems of trade, warfare, and alliance into which they had to seek entry. Control of the region's political economy (and with it the 'monopoly on violence') rested in the collective (though not united) hands of the Lipan Apache, Caddos, Comanches, and Wichitas."[12]

In this contested environment—physical and political—the foraging peoples of South Texas almost had no place. With little to offer in the form of trade with the powerful raiders that galloped through the region in search of foodstuffs, animals, and weaponry, they held a weak negotiating position. Given that "equestrian Indians did everything better than their counterparts who traveled by foot—move, hunt, trade, and wage war," it followed that the pedestrian lifeways of the Payaya made them highly vulnerable to being killed or captured.[13] Perhaps most feared were the Comanche, whose mid-eighteenth-century domain stretched from eastern New Mexico to the rolling plains of Central Texas and down to the Balcones Escarpment at the southern end of the Edwards Plateau, close to San Antonio. Human trafficking was an essential form of capital and status in the Comanche political economy, and although their favorite target was their Apache enemies, they were not particular about whom they swept up, enslaved or sold (or whom they sold them to: other tribes or the Spanish).[14]

The Apache were no more benign. In response to the Comanche's repeated assault on their more sedentary settlements, they sought horses and other plunder wherever they could find them, driving deep into southern Texas and northern Mexico, disrupting putative Spanish "control" of the frontier. Dislocated as well were Coahuiltecan tribal bands, some of whom were already on the move, fleeing northern Mexico to elude the Spaniards' yoke. "They fled the forced labor of Spanish nines and ranches to seek sanctuary in the region of southern Texas," Barr observes, although they did not always find peace. Victimized by the Apache, they also encountered European diseases, "even if they had never had contact with Spaniards," a presaging of the demographic collapse that later resulted in widespread "cultural extinction."[15] These shifting demographics meant that by the early eighteenth century, the region had become "a refuge for an impressively diverse but ravaged congregation of both native and displaced Indians." Their numbers in free fall and their survival in doubt, some of them decided they "might find attractive the food, shelter, and defense of the missions and so offered themselves to the missionaries' teachings and formed mission communities."[16] Faced with the destruction of their way of life, the Payaya made a self-conscious decision to ally with the Spanish, according to Fray Gabriel Vergara, conceding that they did so on their terms. Although he and his fellow missionaries tried to cajole "the Indians with presents as well as by demonstrations of their apostolic zeal, they were not able to persuade the Indians to assemble in the missions," their ministrations seemed to fall on deaf ears. Instead, the "Indians of the Payayas, Aguastayas, and Mezquites . . . were converted for fear of the Apaches."[17]

These conversions were also episodic and negotiated. As Vergara recognized, the Payaya were among those who accepted Christianity and lived in *reducciones* (missions) even as they wandered far afield—attending indigenous sacred ceremonies or foraging for food, drugs, and medicine—acting as "wild Indians" not pacified neophytes. From the missionary point of view, these behaviors were marks of the native people's "free and licentious life," their "depraved habits."[18] A more nuanced reading suggests that these early South Texans, like native peoples across the Spanish northern borderlands, were integrating the missions and missionaries into what archaeologists Tsim D. Schneider and Lee M. Panich call "dynamic indigenous landscapes." This reformulation of Indian agency opens a "window into indigenous people's active negotiation of colonialism and offers a more holistic view on the mission enterprise and its consequences."[19]

There was a limited degree to which the Payaya could remain a distinct people and "indigenize" the Spanish, and that limitation became clear by the mid-eighteenth century: the Payaya *as* Payaya were no more. They could only accommodate so much, as was foreshadowed that day in 1691 when the entrada led by Domingo Terán de los Ríos marched up to San Pedro Springs. That is when a foreign place name subsumed the indigenous; when the written word overlay the spoken. When the generic term "Indios" subsequently became the sole identifier the Spanish employed on mission membership rolls, it stripped away of the names by which native peoples had known themselves. This eradication had immediate implications on the local level, and was also global in its ramifications. At the close of the seventeenth century, writes archaeologist T. N. Campbell, the Payaya and other southern Texas Indians were facing "what most hunting and gathering peoples of the world have had to face: population decline, territorial displacement, segregation and ideological pressure, loss of ethnic identity, and absorption by invading populations."[20]

What endures is an echo of the landscape that the Payaya had created, a flicker of memory carried in the name of a twenty-first-century playground in San Antonio and perhaps the sole surviving word of the Payayan tongue: Yanaguana.

2
Urban Prospect

The Spanish, whether missionary, soldier, or colonist, did not "found" San Antonio. The credit for identifying the fertile river valley's ecological and natural resources and integrating these into local lifeways and far-flung trade networks that crisscrossed the larger region and for which the local springs became one of many nodes, goes to the Payaya and their ancestors. They created the site that the Spanish would adopt for their own, quite different purposes.

It does not diminish the European invaders' influence to indicate that their special contribution was in the *building* of San Antonio. There was nothing simple about their self-appointed task, either. The Spanish would discover that this formidable objective, which took much of the eighteenth century to complete, was more complicated than they had initially imagined. Understandably so, because what they first encountered on the ground fired their imaginations about the nature of the settlement they could construct within the environs of San Pedro Creek and the San Antonio River. Of utmost concern to these early town planners was a ready supply of water, which Spanish visitors assessed in relation to the size and significance of the community it could support. Among the first to do so was Fray Isidro Félix de Espinosa, who accompanied a 1709 expedition to the region with Captain Pedro de Aguirre, commander of the presidio of the Río Grande del Norte, at its head. After marching across the coastal plain, the Spanish worked their way through "a mesquite flat and some holm-oak groves" before coming upon what Espinosa described as an "irrigation ditch, bordered by many trees and with water enough to supply a town." Once the missionary reached its source at San Pedro Springs, however, he broadened his earlier conclusion about the area's potential: "The river [San Pedro Creek], which is formed by this spring, could supply not only a village but a city, which could easily be founded here because of the shallowness of said river."[1] His colleague, Fray Antonio San Buenaventura Olivares, agreed that "an entire province will fit in the said river [valley]."[2]

Captain Domingo Ramón entertained similarly grand visions. On May 14, 1716, he led an entrada that stopped in San Antonio on its way to eastern Texas, there to reassert Spanish claims to that territory in the face of French expansionary threats. "On this day I marched to the northeast seven leagues through mesquite bush with plenty of pasturage. Crossing two dry creeks we reached a water spring on level land, which we named San Pedro," he scribbled in his diary. Like Espinosa, Ramón also identified a number of the site's key environmental features. Noting that the "scenery along the San Antonio River is very beautiful, for there are pecan trees, grape vines, willows, elms and other timbers," his practiced military eye did not miss a critical need for the future movement of people, goods, and services: "We crossed said stream; the water, which was not very deep, reached to our stirrups." That Ramón's boots remained dry signaled that this crossing could serve as an all-essential ford, a transit point that in fact would remain in heavy use until the twentieth century. The Spanish officer also knew that streamflow meant growth and development, and so, more precisely than Espinosa, estimated that there was "sufficient water here for a city of one-quarter league," a bit less than a mile wide.[3]

Two years later, these optimistic reports, when combined with the continuing challenge that France posed to Spanish ambitions in the region, led the viceroy of New Spain to dispatch Martín de Alarcón, the newly minted governor of Texas, to establish a string of communities and missions stretching from San Antonio to East Texas. These "metropolises and capitals," even as they peopled the landscape, would serve as sentry-like settlements, "for observation and defense from seaborne invasions."[4] The viceroy's lofty aspirations ran into recruitment difficulties. Alarcón experienced considerable difficulty in gathering enough volunteers willing to move to the as-yet uncontrolled province. Only seven civilian families made the trek north, in company with a host of missionaries and soldiers, and a large number of livestock, mules, and horses. The expedition trailed into San Antonio on April 25, 1718, and on May 1, the first two formal ceremonies occurred. After consultation with Fray Olivares, Alarcón sited the mission of San Antonio de Valero (now known as the Alamo) immediately to the east of the San Antonio River; four days later, and to the mission's north, he took possession of land adjacent to San Pedro Springs for the civilian settlement he named the Villa de Béxar. A town was born.

Calling the site a town at that date might be an exaggeration, but Alarcón at least had pronounced its presence and proclaimed its broad outlines. His ritualistic actions were according to plan, fulfilling the strictures of the Laws of the Indies (1573). This document, bearing the imprimatur of King Philip II, was a comprehensive compendium of 148 ordinances that wove together recommendations for the selection of sites for a new settlement along with the planning of its built environment and its social and political structures. The Laws, hailed as "the most complete such set of instructions ever to serve as a guideline for the founding and building of towns in the Americas, and in terms of their widespread application and persistence probably the most effective planning documents in the history of mankind," made it clear that Spanish colonization was urban in formulation, communal in import.[5] Ordinance 110, for example, stipulated that leaders such as Martín de Alarcón, having "selected the province, country and area that is the location where the new town is to be built and having taken possession of it," were to act in the following manner: "On arriving at the place where the new settlement is to be founded—which according to our will and disposition shall be one which is vacant and which can be occupied without doing harm to the Indians and natives or with their free consent—a plan for the site is to be made." This entailed dividing the requisite the land into "squares, streets and building lots, using cord and ruler, beginning with the main square from which streets are to run to the gates and principal roads and leaving sufficient open space so that even if the town grows it can always spread in the same manner."[6] The subsequent and rational plan of an interlocking and expanding series of plazas, streets, and houses would produce as well a city whose form "stressed a Christian ideology and a cultural imperialism designed to provide the Spaniard in the

New World" with a recognizable urban environment.[7] Especially for the first generations of Spanish colonists occupying a borderlands outpost, there was no place like home.

As Alarcón took the requisite steps to fulfill his town-planning obligations—including Ordinance 13's demand that he "perform the necessary ceremonies and writs, thus providing public evidence and faithful testimony"—he recognized that a successful civilian settlement required flat, arable, and divisible land with easy access to potable water for household and agricultural purposes. It also had to be readily defensible, hence his decision to locate it on level terrain between the San Antonio River and San Pedro Creek.[8] This initial siting of the town would have mixed results: irrigation proved easier than it would have been around San Pedro Springs, for example, and the soils were richer because they lay within the local floodplain; the latter fact meant that from the very beginning, rampaging floodwaters would be a troubling (and frequent) hazard in San Antonio.

The new town experienced other growing pains during its first decade. The number of Indians—Payayas and others—living within the missions in the San Antonio River valley waxed and waned. To rebuild populations decimated by European diseases and harsh conditions, the friars often used coercive lures. "There are Indians who are hungry, and they accept the faith through the enticement of food," Fray Benito Fernández de Santa Ana advised the viceroy; and there were others who "required weapons of your majesty to bring them into civil society."[9]

Coerced toil defined life at the missions. Rather than teach a higher level of skills, occupations that the Spanish and mestizos (people of mixed Spanish, Indian, and sometimes African heritage) in the villa dominated, the missionaries deployed Indians in "work gangs that provided unskilled manual labor." They did the heavy lifting involved in clearing deep-rooted grasslands and oak woodlands and turning them into arable fields; and then sowed, tended, and harvested the crops. They used rudimentary tools to dig *acequias* (irrigation ditches) out of the stony soil, and lugged sand, water, rock, and timber that would go into the construction of the mission churches, houses, and outer walls. Exploiting the Indians' muscle power had but one goal: "the subsistence, profit and long-range endurance of the missions."[10] That the missions did

Spanish soldiers oversee a group of American Indians digging an acequia. *Courtesy of the Edwards Aquifer Website by Gregg Eckhardt (http://www.edwardsaquifer.net/).*

not endure past the late eighteenth century is linked directly to the high level of Indian desertions and the winnowing population of those who survived this backbreaking existence.

It was not much easier to maintain the number of Spanish residents, a population that across the century remained relatively small and transitory. Some of those who had arrived with Alarcón continued on with him to northeastern Texas, there to plant the Spanish flag and missions. Others, discouraged by the difficulties of life in Béxar, would trickle back to more-established communities along the Rio Grande. In 1724, for example, Alarcon's replacement, Governor Fernando Pérez de Almazán, observed to his superiors that San Antonio was not yet self-sustaining. "It is necessary to seek [recruits] outside this province because here there is no population whatever," but to keep them in Béxar would require an additional incentive "besides their salaries, because of the resignation with which they all come to this country."[11]

Those who remained nonetheless made their way as best they could. The civilians benefitted from conscripted Indian labor extending irrigation ditches from river and creek to bolster local agricultural productivity. The community's clutch of artisans produced goods for sale; as consumers, the population contributed what they could to the town's underdeveloped economy that offered little in the way of manufactured material and services. For all its fluctuations, the population gained in number in part due to uncertainty along the empire's northeastern frontier; when the French attacked Spanish settlements there in 1719, some of the embattled residents retreated to San Antonio. Worried about the defense of this South Texas community, Spanish authorities bolstered its garrison. Although modest in size, roughly 300 people lived along the San Antonio River by the late 1720s; San Antonio had become an established presence on the land.

However fledgling the community may have been, the Lipan Apache found it an irresistible target. They honed in on it because they themselves were locked in a life-and-death struggle with the powerful Comanche who were disrupting the Apaches' once semi-sedentary life on the Plains. To survive, the Apache pressed south and centered their raids on capturing Spanish horses. They were well aware, too, that hostages, livestock, foodstuffs, and weaponry would increase their wealth and trading opportunities, and snatched up these valuable resources whenever and wherever they could. Not surprisingly, they repeatedly attacked the supply trains lumbering between the Rio Grande Valley and San Antonio, and harassed local missions, the town, and the military encampment or presidio. Their successes were not unalloyed. Spanish firepower at times was an effective deterrent: "it is the habit of the Indians to steal, and they do not fail to indulge in thefts, robbing soldiers of their horses if not watchful," one inspection report noted, but "they are generally chastised by the troops for their daring." So were punishing counter-raids. In these circumstances, peace was difficult to maintain. It became even harder after the Spanish allied themselves with the very Lipan Apache who repeatedly had disrupted life in the frontier outpost. At the conclusion of their negotiations in 1749, the combatants buried a live horse and some weapons in a large hole dug in the center of San Antonio's military plaza, a pacific gesture that would ironically generate increased conflict. That was because this new alliance put the Spanish in opposition to the Apaches' long list of enemies, the most feared of whom were the Comanche. Within a decade, these "swift-moving raiders" were mounting attacks on

San Antonio and other Spanish missions and towns, armed forays that continued for the rest of the eighteenth century (and deep into the nineteenth), turning a beleaguered Béxar into the "northeastern lynchpin of defense against the Plains Indians."[12]

The city's increased importance as a defensive bulwark coincided with two pivotal events that occurred in San Antonio in 1731 and that altered the community's physical extent, religious outreach, and social dynamics. Due to Spain's inability to protect its settlements in the war-torn northeastern region, authorities proposed that townspeople, soldiers, and missionaries retreat to San Antonio—a proposition that became a foregone conclusion in 1729 when the military shuttered its presidio there. Within two years, many of those left undefended had pulled up stakes, migrating to Villa de Béxar and points south. Three East Texas missions followed suit. On March 5, 1731, Nuestra Señora de la Purísima Concepción, San Juan Capistrano, and San Francisco de la Espada received lands along the San Antonio River's southern reach. There, they joined Mission San José y San Miguel de Aguayo, founded in 1720, and the upstream Mission San Antonio de Valero (established in 1718), to form a nearly eleven-mile-long stretch of religious institutions. Their combined missionary activity, and increased ranch and agricultural operations, while a boon to the frontier town, added to its complexity. "Depending on circumstances and personalities," historian Félix Almaráz has observed, "throughout the entire colonial period the proximity of church and state institutions in a riparian environment contributed to an atmosphere of cooperation and conflict."[13]

Even more mixed in their impact was the 1731 arrival of a small contingent of settlers from the Canary Islands, locally known as Isleños, who reached the town less than a week after the new missions took formal possession of their riverbank sites. Hoping to bolster the tiny civilian population—there was a total of forty-nine households not affiliated with the presidio or missions—the Spanish crown underwrote the migration of fourteen families from the Canary Islands to San Antonio, the initial wave of a projected four hundred expected to emigrate to Texas. These few were the only ones ever to journey to the province. They received an array of incentives to stay, so many in fact that these newcomers quickly dominated almost every facet of the community. Among the political advantages the crown bestowed on them, writes historian Gerald E. Poyo, was "full control of the town's cabildo, or city council. Ten Isleños received life appointments to govern the new villa, San Fernando de Béxar."[14] This privilege extended to their unchallenged appointment of the sheriff, commissioner of lands, city attorney, and magistrates, and was further reflected in the elevated social status the Canary Islanders asserted. "In Spain's American empire," Poyo observes, "peninsulares [Spanish-born individuals] had always placed themselves at the top of a highly stratified, racially determined social structure." The same was true in San Antonio: the Isleños "looked down on the predominately mixed-blood presidio population that had originally settled the region."[15]

They wielded power accordingly, expropriating access to land and water that had enabled those already established to eke out a less-than-robust existence in the town. Within days of their arrival, the presidio's commander had granted the Isleños ownership of the arable land and related acequias south of the fort that since 1718 his soldiers had cleared, worked, and irrigated. Not satisfied with this expropriation, the immigrants then "claimed rights to virtually all the nonmission land west of the San Antonio River and sought to monopolize water from the river itself and the San Pedro Creek."[16] Within a decade, their control was complete and had even informed their presumption that they were the community's first settlers, that it was they who had pacified the area. Those soldiers and civilians who had arrived more than a decade earlier rebutted this assertion, alleging in a 1745 petition that the Canary Islanders were so "conceited by the title of [original] settlers" that their wish to be "the only [settlers] of that land" thereby "scorning "the *agregados* who at no cost to His Majesty were and are the true and most ancient settlers and conquerors of that land."[17] Yet for all their assumed power, the Canary Islanders remained a querulous lot. "The fourteen families from the Canary Islands complain against the reverend fathers of the five missions," a contemporary wrote, "against the Indians who resided therein, against the captain of the presidio, and against the other forty-nine families settled there, so that it seems they desire to be left alone in undisputed possession. Perhaps even they may not find enough room in the vast area of the entire province."[18]

By the end of the eighteenth century, however, the realities of life on the empire's periphery had worn down the Isleños' hard-edged arrogance. They shared in the town's rough conditions,

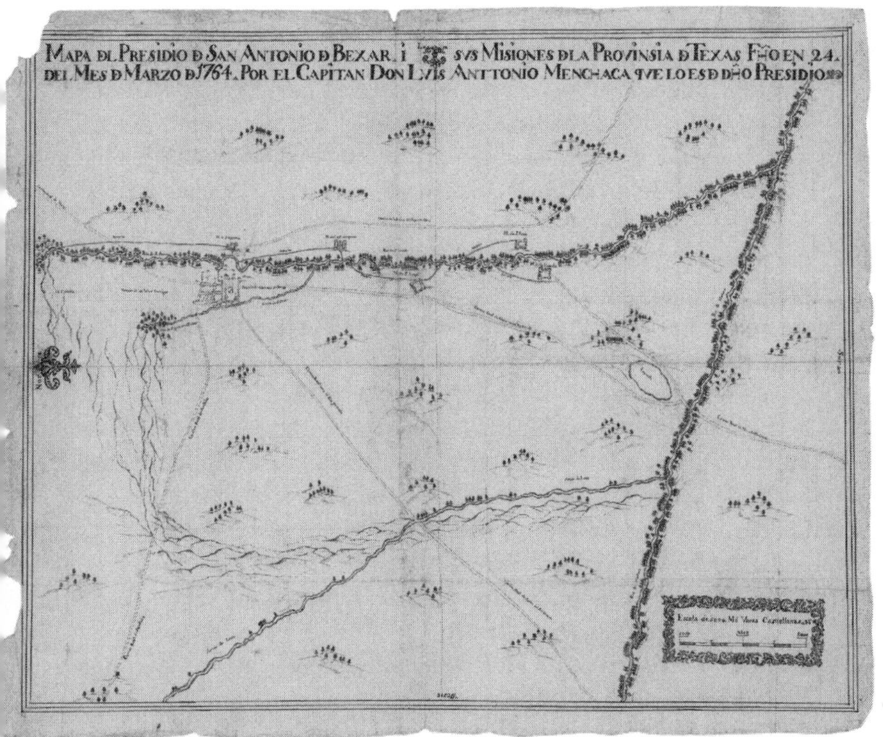

1764 map of the presidio of San Antonio de Béxar. Created by Luis Antonio Menchaca. *Courtesy of the John Carter Brown Library at Brown University.*

which caught the eye of the waspish Fray Juan Augustín Morfi, who toured San Antonio in the late 1770s—and found it wanting. The built environment, he wrote, consisted of "fifty-nine houses of stone and mud and seventy nine of wood, but all poorly built, without preconceived plan, so that the whole resembles more a poor village than a villa, [a] capital of so pleasing a province."[19] The inhabitants of this mud-caked town were in even worse shape, the scandalized cleric observed; the entire population, Isleños or otherwise, were "indolent and given to vice, and do not deserve the blessings of the land."[20]

For all his disdain, Morfi was not wrong to conflate the Canary Islanders with those who also called San Antonio home. Their one-time exclusivity slowly gave way before an interrelated series of inescapable pressures. Intermarriage, and a corresponding spinning out of more complex kinship networks, was one factor that over a couple of generations had diminished their attempts to maintain their distinctiveness. "Social integration also had a spatial dimension," Gerald Poyo argues. "In the early years of the settlement Isleños lived around the Plaza de las Islas, while the military families lived either in the presidio or adjacent to it," but this physical separation became much harder to maintain as the community filled in the available open space and then pressed outward. The urban economy also proved integrative: although it took fifty years for presidial farmers to recover their usurped rights to land and water, once reclaimed the Isleños' preeminence diminished. That decline came paired with their belated recognition that ranching—an occupation that soldiers and civilians had taken up when they lost land and water rights—required less start-up capital and produced greater opportunities than did subsistence agriculture. This convergence of economic interests across class barriers and racial lines, when linked with changing social mores and a more inclusive political structure, helped create "a unifying cultural experience that promoted a common Tejano identify."[21] An outcome that had been unimaginable forty years earlier.

This newly forged social type, which in San Antonio was identified as *Bexareño*, was perhaps most intriguingly manifest in the community's protest against the imposition of greater royal control over provincial affairs. At the same time that the English colonists in eastern North America challenged the right of their king and colonial governors to determine their political future, and in response to the so-called Bourbon reforms that led Spain to impose tighter regulations on colonial governance and economies, those living in San Antonio pushed hard to maintain local autonomy. "Whenever Bexareños perceived a threat to their livelihood," historian Poyo argues, "they closed ranks politically to defend their community and shared destiny."[22] As this perception of a shared set of common interests grew, it set the stage for a movement that would contribute to the establishment of an independent Mexican nation-state and the subsequent Texas Revolution. As it had been the case during the eighteenth century, political tumult, armed struggle, and social upheaval would mark San Antonio in the nineteenth.

3
Revolutionary Space

San Antonio was contested ground by the middle of the eighteenth century. Its mission Indians, soldiers, and the larger Tejano population had become battle-hardened—and battle-weary. If they were not fending off swift raids that the Apaches and Comanches launched against the town, missions and ranches, they were marching off on punitive counter-strikes. The possibility of either scenario disrupting the rhythms of the agrarian cycle and local trade with distant markets and fairs gave them little down time. In this community, "with more battles fought in or around it than any other in what is now the United States," almost constant conflict made an already rigorous and uncertain frontier life even more so.[1]

These perils of war declined somewhat in the 1780s after the Spanish government and the Comanches established a relatively stable peace agreement. A sizeable trade then emerged between Comanchería, the tribal territory, and San Antonio then emerged. "The Indians brought hides, tallow, meat and sometimes horses and even captives to exchange for textiles, clothing, and ornaments," though apparently the most prized object was "the dark, sweet piloncillos, the sugar cones of which there could be never enough to sate Indian appetites."[2] So extensive was the import-export business that the community built a thirty-six-room lodge in which to house their Indian partners, and the site quickly became a profit center. There, goods and services were bought, traded, and sold, a prosperous exchange that local merchants carried south for resale, material that soon constituted "the bulk of San Antonio's growing trade with Saltillo, which was its principle link to the economy of New Spain."[3]

The stabilization of the local marketplace came at the same time that one of Béxar's principal institutions—the missions—was destabilized. Their success as independent and insular communities had never been as pronounced as the Franciscans had hoped, in good measure because European-introduced diseases decimated Indian populations and compelled the missions to recruit neophytes from across South Texas, through means peaceful and violent. Their agrarian economy was also in competition with civilian farming and ranching, a competition that intensified as peacetime trade with the Comanches accelerated. The imperial government also conceived of the missions, once the mainstay of its pacification policies, to be a cost it no longer wished to absorb. The Franciscan Order initially resisted this conclusion, but the local council, responding to demands of landless residents, among them mestizos, Indians, and Spanish seeking access to arable land, ultimately agreed with these civilian petitioners. Local leaders argued that decommissioning Mission San Antonio de Valero, which controlled the "principal irrigation works from the spring of the San Antonio River, with much more irrigated land than any of the other missions, or even this town," would facilitate the town's growth and development.[4] In 1793, Spain announced that it was secularizing the mission and then set about distributing its property. So as "to prevent disputes and avoid misunderstandings," the formal announcement declared, "we shall proceed to partition the land by drawing lots, and each grantee must be contented with the land that falls to him, whether it is plowed land or timber land or if it is provided with irrigation or lateral conduits."[5] Among the beneficiaries were "fourteen Indian families, forty-two Adaesano [those who had resettled in San Antonio following the collapse of the eastern Texas settlements] heads of households, and fourteen others."[6]

The transition was less complete and equitable at the other missions, all of which were at least partially secularized in the spring of 1794. At Mission San José, the Franciscan priest remained to serve the local population's spiritual needs although its lands were distributed. Some of its property went to its resident Indians, while townspeople snapped up the remaining acreage. Mission Concepción, which lost its independent status and became a sub-mission to San José, also continued to enjoy Franciscan spiritual support. Its lands were distributed in part to the Indians still living on site,

Mission San José. *Library of Congress Prints and Photographs Division, Washington, D.C.*

The acequia at Mission Espada. *Library of Congress Prints and Photographs Division, Washington, D.C.*

while the government set aside some lots for community use and others for its taxing purposes. The much smaller Mission San Juan Capistrano, which had never flourished on its smaller amount of land, shut down in June 1794. That same month, lands associated with San Francisco de la Espada, the farthest south of the missions, were also distributed to its tiny band of resident Indians, many of whom were too old or incapacitated to farm and thus were formally encouraged to rent out their parcels to other converts or local Tejanos. Full and final secularization came in 1823–24 following the struggle for Mexican independence. (Mexican Revolution usually refers to events of the 1910s), with the pastor of San Fernando Church assuming spiritual and administrative supervision of the missions.[7] By this date, the missions' walls, churches, and other structures had begun to fall apart, a physical deterioration that would not be arrested until the 1930s when these built landscapes became the focus of an emerging preservation movement dedicated to restoring (and appropriating) the city's Spanish heritage. Many of the original owners of the land that the government had distributed in the 1790s had lost their hold over it by the time of the Mexican wars of independence, perhaps a sign of the degree to which the former mission Indians and their progeny had, mostly through intermarriage, been absorbed into the larger community. A hopeful reading of this dynamic suggests that among its outcomes was a strengthening of bonds between the Indians, Tejanos, and other residents of mixed parentage, "resulting in lasting Indian native traditions in the town and a progressively integrated Indian-Hispanic society in San Fernando de Béxar."[8]

The importance of large-scale geopolitical events, then in considerable flux, would dwarf these internal realignments. San Antonio and its small population, which in the 1790s amounted to roughly 1,300 souls, and which by 1820 had grown to about 2,000, had no control over these shifting realities that would turn their world upside down. The first of a series of dramatic alterations involved Louisiana. Spain had gained control of it from France in 1764, easing Spanish concerns about defending its eastern frontier. Until, that is, the westward ambitions of the United States emerged at the last years of the 1790s, at which point Spain closed New Orleans to American travelers and commercial traders. This new imperial power might have given Spain more pause had it not been under dire pressure at home during the Napoleonic Wars, in the midst of which Madrid secretly returned the vast territory to France via the Third Treaty of San Ildefenso of 1800. Within three years, Napoleon sold Louisiana to the United States, catching Spain off guard and feeding America's insatiable land hunger. Once a buffer against the expansionist designs of Spain's rivals, Louisiana suddenly had become a staging area for an even more aggressive adversary. As a result, San Antonio returned to its role as a Spanish redoubt: the local garrison of 100 soldiers swelled in size, with the viceroy dispatching another 450 to hold the frontier town.[9]

That show of force came at the same time that a weakened Spain began to implode. With its domestic economy in freefall, and its agricultural production in decline, the government demanded an increase in colonial revenues, and appropriated the church's charitable funds, which led the church to call in outstanding loans and other obligations—further depressing Spain's economy. Napoleon could not have picked a better time to invade the Iberian Peninsula, and in short order had placed his brother on the Spanish throne. What happened in Europe did not stay there. Elites in Mexico City sought to stabilize the colonial viceroyalty system of governance; others pressed for a radical rearrangement of power. In 1810, a rebellion broke out in Guanajuato, with Father Miguel Hidalgo y Costilla leading a mixed-race army, the composition of which alone frightened some potential sympathizers enough to ally with the viceregal forces. The bloody insurrection, which over the next five years evolved into "a mestizo movement for independence," was stamped out in November 1815.[10]

The reverberations in San Antonio, though far from the center of action, were immediate. In January 1811, rumors reached the city that seemed to suggest that the royalists were being routed. This news led a retired Spanish army officer, Juan Bautista de las Casas, to conspire with disgruntled members of the local militia to strike against Governor Manuel Salcedo. Over the next thirty-nine days, after solidifying control in San Antonio, Casas sent an expedition to Nacogdoches to establish rebel rule there, and was acclaimed the interim governor of the province of Texas, only to lose to a counterrevolutionary force comprising some of his one-time allies. Imprisoned in Mexico, Casas was shot and beheaded on August 3, 1811; the government put his skull on display in San Antonio.[11]

The gruesome warning did not restrain the revolutionary impulses surging through San Antonio and the rest of the province. One year later, in August 1812, the so-called Republican Army of the North, under the command of Hidalgo's supporter Bernardo Gutierrez de Lara, and several American filibusters, swept into Texas from the northeast. This well-armed contingent overwhelmed the garrison at Nacogdoches and by November had gained control of the lightly defended town of La Bahía (present-day Goliad, near the Gulf Coast). Hoping to check their progress, Governor Manuel Salcedo mounted a siege of the town; when that failed, he and his forces retreated to a site near the confluence of Rosillo and Salado Creeks, southeast of San Antonio. On March 29, 1813, the battle was joined and, in less than an hour, the royalists were routed. Two days later, San Antonio capitulated to the Republican Army of the North. Its vengeance-seeking leadership, perhaps as a reprisal for Casas's beheading a year earlier, brutally executed Salcedo and other royalist military officers—they slit their throats and left them exposed where they fell.[12]

The cycle of carnage continued. As the new revolutionary capital, San Antonio was awash with rumors and backstabbing; leadership positions seesawed, soldiers lacked pay and their morale plummeted. Those who had lost in this round of fighting lived on and conspired. Then in August, a new royalist army, 1,800 men under the command of General Joaquín de Arredondo, pushed north from the Rio Grande and met the republican army along what today is known as Old Pleasanton Road, close to the border of Bexar and Atascosa Counties, a site that is twenty-two miles to San Antonio's south. In what is believed to be "the bloodiest battle every fought on Texas soil," Arredondo's forces slaughtered the Republicans, killing an estimated 1,300 of the 1,400-man army. The bloodbath led the victor to crow: "The ever victorious and invincible arms of our Sovereign, aided by the powerful hand of the god of war, have gained the most complete and decisive victory over the base and perfidious rabble commanded by certain vile assassins ridiculously styled as general and commanders."[13] The rebel dead he left unburied, as carrion and as a message; the corpses would rot and bleach in the hot South Texas sun for nine years before their bones were interred. Among those commended in Arredondo's official report, and who later adopted his commander's zeal, was Lieutenant Antonio López de Santa Anna. He was not alone: when news of the royalist victory reached San Antonio, where twenty-three years later Santa Anna would mount his historic siege of the Alamo, royalists murdered republicans before they could flee, and the imperial army, once it gained control of the town, chased down and killed more.[14]

No stranger to the ferocity of warfare, an ongoing consequence of San Antonio's role as the vital hub of the northern province, the town nonetheless had never witnessed anything like the violence that erupted between 1811 and 1813. Although Spain may have prevailed during this two-year-long struggle, its victory was Pyrrhic. With neighbor turning against neighbor, the city's social fabric was rent, and its population shrank. San Antonio's economy was also in tatters—the undefended town became "subject to constant Indian depredations; [and its] ranches were deserted and herds either destroyed or taken to safety in the interior"—leaving the community in disarray.[15] Knitting it back together would take time, a patience that was a measure of the Tejanos' pragmatic approach to managing turmoil. Pragmatism could only achieve so much, however, when confronted with outside pressures and internal unrest that over the next twenty-five years led to the collapse of Spanish rule, the birth of the Mexican nation-state, and yet another revolt that ended with a new power in control of Texas and its most important city.

One of those whose life story dramatically reflects these radical shifts in the early nineteenth-century community was José Francisco Ruiz. Born in San Antonio in 1783 and educated there by his Spanish-born father, in 1803 Ruiz accepted an appointment as the frontier town's schoolmaster, holding his first classes in his home fronting Military Plaza. Like many of his peers, he sympathized with the republican cause, but unlike them, he managed to survive the Medina bloodbath, and with his nephew José Antonio Navarro, found sanctuary with the Comanches and in Louisiana. His 1818 report on Comanche lifeways remains an influential analysis of their communal interactions and inter-tribal alliances, and was also the means to his return home. In 1821, the newly independent nation of Mexico offered him a pardon if he would serve as its peace commissioner to the Comanches and Lipan Apaches (with whom he had also had considerable interactions). Ruiz agreed, and one year later the Apaches signed a treaty in Mexico City that Ruiz had helped broker. For the next decade, he continued to be the republic's emissary to the northern tribes, including the Wichitas and Comanches. The Shawnee also knew about his integrity. As Thomas McKinney of Nacogdoches wrote Stephen F. Austin: "the Shawnees have talked of going for some time to see a big man in St. Antonio who they say is a good man no lie and a good friend to the Indians, Ruis."[16] As another revolt erupted in 1836, once again Ruiz was at its center: he was one of two Tejanos to sign the Republic of Texas' Declaration of Independence (the other was his nephew, Navarro). Ever alert to the tremors of war, and the resultant costs that Tejanos might be compelled to pay, he urged family and friends in San Antonio to tread carefully, writing in December 1836: "Under no circumstance take sides against the Texans, for only God will return the territory of Texas to the Mexican government."[17]

The devastating political tumult had its environmental analog. Even as contending armies and factions fought over the city, nature reminded everyone of its preeminence. The flood of 1819, for example, highlighted some of the steep disadvantages under which San Antonio had labored at the start of its second hundred years. That July, one day after Governor Antonio Martínez had hastened to assure his superior, Commandant General Arredondo, that "no special incident has occurred in this province under my command," a storm blew over the San Antonio River watershed, causing a flood "so terrible that no object could resist its fury." The churning river blew out of its banks, raced west to join with an already swollen San Pedro Creek, and the combined erosive force wreaked havoc. "After the stock in the *potrero* disappeared," Martínez wrote, "the water began to dislodge and carry away several *jacales* [houses] from their locations, and the people who lived in them were seen as they were carried down the river by the strong current." It was "impossible to give immediate aid to the miserable souls who struggled against death because no one could do anything except look out for himself." The "indescribable violence" took out the city's main bridge, destroyed government offices and documents, stripped people of their food and clothing, swamped the quartermaster's granary, and left behind sodden cornfields, gardens, and grasslands. Without the resources needed to lift the citizenry's burdens, Martínez begged his superiors to respond with alacrity: "I solicit your mercy in the name of the unfortunate people who appeal to you as the father of these provinces" to alleviate "their misery insofar as possible, [and] to take the most effective measures that your meager resources will allow."[18]

The flood, which exposed the hazardous nature of the city's siting, also revealed that New Spain itself was foundering. Arredondo apparently sent no funds or supplies, as there was nothing to dispatch to the shattered town on the frontier's edge. Martínez's subsequent letters, which sought more soldiers and weapons to fend off Indian raids and to help counter the revolutionary fervor that had not died down, went unanswered as well. His "piteous appeals for help" fell on deaf ears, "for neither the viceroy nor the commandant general could render effective aid." The Spanish treasury was exhausted and the royalist cause endangered. As the local *ayuntamiento* (town council) reported one year after the damaging floodwaters had receded, the community and province was racing, "at an amazing rate, towards ruin and destruction."[19]

Little changed in the wake of the March 1823 abdication of Emperor Agustín de Iturbide and the subsequent establishment of the Mexican republic. The new government reorganized provincial affairs, stripping San Antonio of its status as capital of Texas and merging the province with Coahuila. The state legislature was located in Saltillo, many hard-days travel away. This political union troubled many in San Antonio, writes historian Jesús F. de la Teja: "having acquired a taste for self-rule, Tejanos were reluctant give up to the provincial deputation to a legislature in far-off Saltillo."[20] That the new capital was 345 miles away, and required a rugged journey across a hot and hostile terrain, did not increase the chances that San Antonio's pleas for support would be sympathetically received, argued Erasmo Seguín, Texas's representative to the Second Constituent Congress, which drafted the new country's federal constitution. The "inhabitants of Coahuila," he wrote home in December 1823, "will not help us with money to get us out of the difficulties in which we find ourselves, nor to preserve our lives and vital interests." His words echo the hard-won experience of someone on the other side of the political divide—Antonio Martínez, the last Spanish governor of Texas. Even a radical shift in government could not alter San Antonio's on-the-ground realities.[21]

Strikingly, Martínez and the nationalists, whose triumph sent him back to Spain in 1822, also agreed on what had the best chance of energizing San Antonio's laggard economy and bolstering its political influence. In 1821, in hopes of increasing the number of people able to defend colonial settlements from the Comanches and of boosting commercial and agricultural productivity, Martínez reluctantly approved Moses Austin's petition to transplant three hundred Anglo American settlers to the province. Some of the new arrivals were slave owners and brought their human chattel and cottonseeds to replicate the plantation system they had left behind. To protect these wealthy new slave-owning settlers, San Antonio's town council and its representatives petitioned a series of national assemblies, pressing the new nation not to abolish slavery. The debate was heated, and the final resolution, after an initial abolitionist clause passed, was diluted such that the slave trade alone was prohibited, "leaving unclear whether the prohibition extended to owners bringing slaves for their own use."[22]

The land rush was on. Organizers of American colonists such as Moses Austin's son, Stephen F. Austin, worked in close collaboration with the San Antonio *ayuntamiento*, in locating new settlements close enough to the city to aid its growth and defense. "I cannot help seeing advantages which to my way of thinking would result if we admitted honest, hard-working people," observed one Bexareño, "regardless of what country they come . . . even hell itself."[23] Not everyone concurred that Hades was a decent source of immigrants. For them, the new wave of immigration was hardly a unilateral social good. As squatters proliferated, tensions mounted over property boundaries and ownership. Still, land speculation was rampant because the profits were great; for San Antonians long used to being land rich but cash poor, the prospects of building up a nest egg from these speculative ventures was enticing. The central government disagreed, and beginning in 1829 tried to control unchecked in-migration in two ways: abolishing slavery and prohibiting Anglo Americans from entering Texas. San Antonio's representatives to the state and national legislative bodies protested the implementation of these regulatory constraints, protests that merged with a larger argument about the new country's governing structure. "The political situation in Texas became increasingly complicated," historian Jesús F. de la Teja argues, as the "issues of slavery and immigration became intertwined with the national debate over the need to strengthen the country's government."[24]

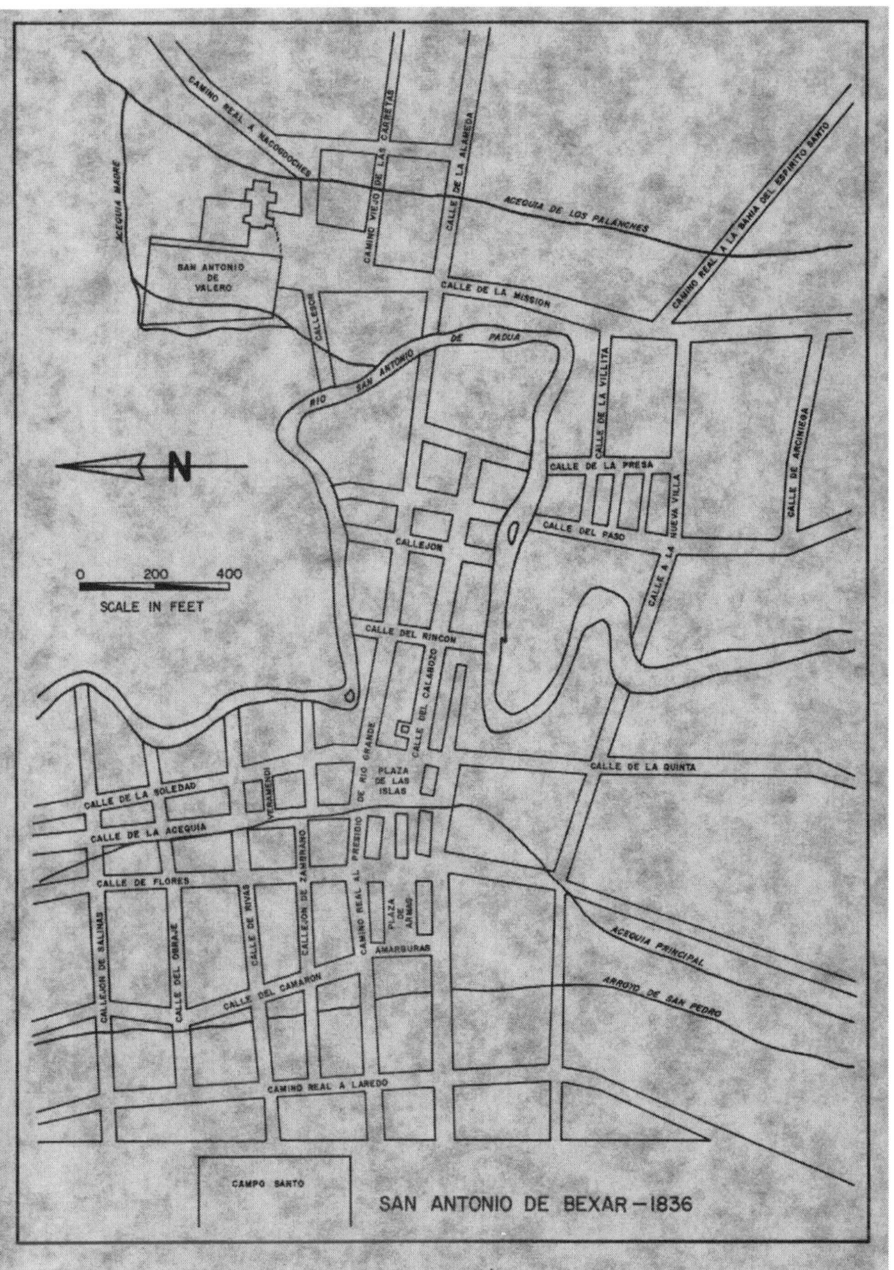

Map of San Antonio de Béxar, 1836: *From Texas State Publications, provided by UNT Libraries Government Documents Department to the Portal to Texas History (https://texashistory.unt.edu/), a digital repository hosted by the UNT Libraries.*

Those advocating for the centralization of power demanded greater control over immigration and the dispersal of public lands; those favoring a more decentralized government believed that a liberal, federal structure best served their interests. Many of San Antonio's Tejano population were in the latter camp, as was true of their Texian (Anglo) neighbors. Mexican centralists carried the day in 1834, when the one-time liberal, President Antonio López de Santa Anna, converted to their cause and moved to squelch regional opposition. Texas, once again, was in the crosshairs.

The Tejanos of San Antonio were caught in the middle of the unfolding conflict. As opposed as they were to Santa Anna's dictatorial rule, they had no love for the Anglo Americans' bellicose rhetoric and aggressive conviction that outright independence from Mexico was the only solution to the constitutional crisis. The Texians precipitated war on October 2, 1835, when they defeated units of the Mexican army near Gonzales; ten days later, they laid siege to San Antonio and the forces there under the command of Martín Perfecto de Cos. As General Cos fortified the Alamo and other defensive structures, the Texians debated strategy and, for the next two months, the combatants tested one another's strength in a series of inconclusive skirmishes on the city's outskirts. Finally, in early December, the battle concentrated within San Antonio proper, with trench construction, hand-to-hand combat, and cannon fire damaging buildings and streets in close proximity to the Alamo. The home of Don Juan Antonio Chávez was among those ruined: "the walls had wide rents in them made by the cannon shot," he recalled. "In them were many bullet holes and marks and the doors and windows had been pierced and riddled."[25] Unlike the next year's explosive assaults on that soon-to-be sacred site, this first battle for the Alamo ended anti-climatically: Cos surrendered on the condition that he and his soldiers receive safe passage south, which they received. Santa Anna would offer no such quarter to the Texian army that, four months later, he surrounded in the former mission, a refusal that would give the second battle of the Alamo its legendary cast.[26]

The ferocious clash in 1836 between Santa Anna's 4,000-man army and the 182 individuals holed up in the Alamo might have been avoided. When a Texian lookout atop the town's church spotted a cloud of dust rising to the southeast, and warned that the Mexican army was advancing on the city, Colonel William Barret Travis, who commanded the small rebel force, could have ordered his men to fall back until they met up with the main body of the Texian army. He may have decided not to because he expected that the revolutionary government would send reinforcements. When none came, he might have slipped away: in retrospect, confronting Santa Anna at full force, as happened later at the battle of San Jacinto, and at which the Texians crushed the Mexican army, would have made strategic sense. Indeed, José María Rodríquez remembers his father, who would fight in a Tejano unit under Sam Houston, twice advising Travis to do exactly that. Travis instead countered that holding San Antonio was of overwhelming military importance. As he wrote in one of his last letters before the final battle: "The power of Santa Anna is to be met here or in the colonies; we had better meet them here, than to suffer a war of desolation to rage our settlements." He therefore consolidated his men inside the now-fortified former mission, telling the senior Rodríquez: "we have made up our minds to die at the Alamo fighting for Texas."[27]

For his part, Santa Anna also knew how critical it was to capture San Antonio, a lesson he had learned in 1813, when as a Lieutenant he had fought under General Arredondo in his brutal campaign to regain the city from the Republican army. Still, with the odds overwhelmingly on his side, Santa Anna might have mounted an ironclad siege, shutting off food and water and letting hunger, thirst, and disease incapacitate those who chose to remain inside the Alamo. That did not exactly happen, either. On February 23, after the Texians refused to surrender, Mexican artillery opened fire, a steady cannonade that hammered the walls for the next twelve days. Although his subordinates urged continuation of the siege, the ever-hasty Santa Anna ordered an all-out assault on the early morning of March 6. The ninety-minute battle could be seen from miles away. Young Rodríquez, who with his mother and siblings had decamped for a distant ranch, stood on the roof of their refuge; from it, he could "see the flash of guns and hear the booming of the cannon." Eulalila Yorba, who stayed in the smoke-filled city, huddling at her parish priest's home: "There was nothing to impede the view from the priest's house, although I wish there was." She watched in horror as cannon and muskets let loose volley after volley. "The din was indescribable. It did not seem as if a mouse could live in a building so shot at and riddled as the Alamo that morning." Even when the reverend father drew her away from the window, she "could still hear the shouts and yells and booming of the brass cannon [that] shook the priest's house and rattled the window panes."[28] The ferocity of the assault, like the courage and heroism of all those who endured it, ended in deafening silence.[29]

The Alamo's shock waves, when combined with the mind-numbing news of the indiscriminate murder of 342 men who had surrendered to Santa Anna's forces at Goliad, would awaken the Texians to their "perilous situation," observes historian Stephen Hardin, finally forcing them to recognize "that the war was far from over and that they must unite to fight or lose all." Nothing about this awakening was straightforward, but at San Jacinto (near present-day Houston) on April 21, 1836, the lucky Texian army managed to overwhelm and defeat Santa Anna's forces. Captured the next day, the Mexican president was treated with far greater generosity than he had shown those whom he had vanquished. However accidental this victory—"San Jacinto was not so much a battle that Houston won, but rather one that Santa Anna squandered"—its impact on San Antonio and its Tejano residents was of considerable consequence.[30] The creation story of the ensuing Republic of Texas would become wrapped in mythic garb, woven from the bloody cloth of the Alamo martyrs. Their martyrdom went national following Texas's annexation to the United States in 1845 and was then fused with an earlier revolutionary ethos, the "Spirit of '76." It has loomed large ever since, dominating the Anglo American master narrative of how the west was won, how Manifest Destiny played out in and beyond Texas.

The first act of this drama, tinged with racism, was aimed at Tejanos who bore arms against Santa Anna; whisper campaigns among the army's Anglo leadership questioned these men's loyalty to their shared cause. After the war for independence, Anglo-controlled communities in eastern Texas petitioned the new republic's legislature to disenfranchise all Tejanos, veteran or otherwise. Those who fought were routinely denied service pensions. In Tejano towns in South Texas, Anglos pursued a brutal policy of occupation.[31] American settlers swept into La Bahía, razing its buildings and forcing out its entire population. In Victoria, "loyal Mexican families were driven from the homes, their treasures, their cattle and horses and

Juan Seguín. *Courtesy State Preservation Board, Austin, Texas; Original Artist: Wright, Thomas Jefferson / 1798–1846; Photographer: Perry Huston, 7/28/95, post conservation.*

their lands, by an army of reckless, war-crazy people." In San Antonio, dispossession and alienation was to a lesser degree also local Tejanos' lot. José Antonio Navarro was among local Tejanos who struck back in defense of "the old settlers of Texas, who are wantonly attacked by those who come into the country . . . to snatch from the fruits acquired by perseverance and fortitude."[32]

Juan Seguín, who fought at San Jacinto, was the only Tejano to serve in the congress of the new republic and in 1841 was elected mayor of San Antonio. He struggled to protect his people. "The American straggling adventurers," he lamented, "were already working their dark intrigues against the native families, whose only crime was, that they owned large tracts of land and desirable property." Like Navarro, he interceded when and where he could, but Seguín knew who and what he was up against: he and his compatriots were "exposed to the assaults of foreigners, who, on the pretext that they were Mexican, treated them worse than brutes." Through intimidation, violence, and discrimination, Tejanos' political rights, legal standing, and social status vanished. Seguín's public career provides a telling example of the speed and completeness with which they lost power and place: when he resigned as mayor in 1842, he would be the last Spanish-surnamed individual to hold that position for the next 140 years. Put another way, by the advent of the U.S.–Mexico War in 1846, Tejanos were fast becoming strangers in their own land; Béxareños, exiles at home.[33]

4

FORCES OF AMERICANIZATION

Looking out the windows of the Bexar County Courthouse, William Giles Martin Samuel observed the passing parade and then painted what he saw in a set of four views of Main Plaza, the city's central spiritual, commercial, and political hub since the Spanish laid it out in 1718. Although he may have produced these paintings after the Civil War, Samuel identified them as capturing the city's street life and built landscape in 1849. His is a beguiling record of the community's pre-railroad bustle, the daily rounds and daily hustles that defined San Antonians' interactions with one another and their economic activities. Samuel framed each scene by its compass position: north, east, south, and west. The western perspective is perhaps the most realized and certainly the most energetic—a reflection of the many ways contemporaries utilized the open-spaced, hardened-dirt plaza. Slightly off-center and serving as the apex of the image is San Fernando Cathedral, through whose eastern-facing door the pious (and occasionally the impious) would flow, marking major and minor moments in the Catholic liturgical calendar. In the middle and foregrounds of Samuel's painting is the where the greatest action lies. A stagecoach rolls north across the plaza, while horses, dogs, and cattle, burros and carts dot the canvas. A vaquero twirls a riata above a cow-at-large. Four dogs wrestle in the dust. Wood-laden mules haul in kindling and split logs from the forested hills to the city's north and west; ox-carts lug in heavier loads of hides for local tanneries. Captivatingly, a lone horseman, with a large blue saddlebag, canters south. Even those on the painting's margins—whether sitting, standing, or leaning against a post—offer an important counterpoint to Samuel's primary message: San Antonio was on the move.

William Giles Martin Samuel's *West Side Main Plaza San Antonio Texas 1849* captures the importance of Main Plaza to the daily life of San Antonio during a time of tremendous change. *Courtesy of Bexar County Commissioners Court.*

The source of that movement, the painter implies by the placement of the most colorful, eye-catching object in the center background of the image, is the nation-state that now controlled the city: a wind-rippled Stars and Stripes flutters over all. In 1849, the flag's presence was an everyday reminder of the region's rapid transition from Mexican province to an independent republic to a state within the American Union. In 1840, the last direct, and very bloody, battle between the city's residents and Comanche warriors had erupted in and around Military Plaza; known as the Council House fight, dozens of Comanche were killed, triggering vicious reprisals west of the city. Two years later, in the spring and fall of 1842, the final (and brief) Mexican army occupations of San Antonio had occurred. By choosing 1849 as the date of his painting of Main Plaza, and emphasizing its quiet ordinariness, Samuel underscored what peace looked like in a city that was only one year removed from the end of the United States' war with Mexico.

Not that this was Samuel's point, but San Antonio's proximity to the Rio Grande Valley, the still-contested southern border of the new state of Texas, had figured importantly in President Polk's larger decision to provoke hostilities to settle its claims to the borderlands region. Along the way, and not incidentally, the expansionist nation would also conquer territory that would become the states of New Mexico, Arizona, and California, and portions of the future states of Colorado, Utah, and Nevada.[1] For its part, the Alamo City became a staging area for some of these military operations. In June 1846, having received orders to lead an expedition from San Antonio into Mexico, there to capture Chihuahua, General John E. Wool and his contingent of cavalry and troops headed south from the Ohio River Valley; they rendezvoused in New Orleans. From there, mounted units headed cross-country to San Antonio while the foot soldiers caught transport to Port Lavaca on the Gulf Coast before marching to the city. Nature complicated their journey: Heavy showers in July and August flooded the region, turning the Lavaca–San Antonio road into a mud-sucking mess. Once they reached San Antonio, they found, much to Wool's dismay, that the city was not ready for the influx of men and material. Troops had to be stationed in different locations: the cavalry stabled south of the town, near Mission Concepción, while most of Wool's forces, officially called the Centre Division, bivouacked at what came to be called Camp Crockett, close to the headwaters of the San Antonio River. Wool discovered there were not enough provisions readily available to feed men and animals. There were also not sufficient quantities of horses, wagons, or carts to meet transportation needs; among the city's teamsters ready to pitch in was William Samuel, amateur artist. To resolve these operational conundrums required the general's focus for the next couple of months. Even as the situation improved, when he and his command of 1,400 men and 113 wagons headed southwest toward the Rio Grande in late September, they left behind several regiments.[2]

Once in Mexico, Wool discovered another predictable problem from which San Antonio had long suffered, the difficulty of transportation—every mile he and his men marched into the Mexican interior was another mile army supply trains would have to cover. "The bases were too remote from the supplies, and the lines of operation much too long," observed George W. Hughes, Wool's Chief of Topographical Staff. The U.S. Army's logistical problems may have been worse than those Napoleon suffered during his ill-fated invasion of Russia. The "distance from San Antonio to the city of Mexico, by the way of Chihuahua, over any known route practicable for an army carrying with it artillery and munitions of war, cannot be less than two thousand miles," an assessment that affirmed as well the chief difficulties San Antonio had suffered from during its years as a Spanish and Mexican outpost. The land traversed posed additional and unforeseen complications: "long destitute of water and sustenance," Hughes wrote, these rugged terrains are "mere desert wastes."[3]

Whatever its logistical shortcomings and its off-the-map quality (on the march south, Hughes related, "we were almost literally compelled to grope our way, and, like a ship at sea, to determine our positions by astronomical observations"), San Antonio had its magnetic moments. This was particularly so of Main Plaza and its environs, which became the soldiers' favorite location for release and recreation. Saloons, gambling dens, and dance halls were packed; the men reveled in what they perceived to be the exotic and erotic nature of the fandango, a Spanish courtship dance. One

Fandango, c. 1858. *Frank Leslie's Illustrated Newspaper.* This image captures a popular form of entertainment in San Antonio in the eighteenth and nineteenth centuries, an evening of dancing, drinking, and eating. Local elites and religious authorities believed fandangos were licentious, and officials tightly regulated them—which did little to stem the dance's popularity.

of the entranced, Sam Chamberlain, a member of the Second Dragoons (a type of cavalry unit), wrote a tell-all memoir after the war, *My Confessions*, in which he regaled his readers with tales of his full-on immersion in the city's nightlife. Sauntering into the Bexar Exchange, a notorious dive, Chamberlain surveyed the raucous scene: the bar room was crowded with volunteers, regulars, Texas Rangers, a few Delaware Indians, and Mexicans. The Rangers, he wrote admiringly, "were the scouts of our Army and a more reckless, devil-may-care looking set, it would be impossible to find this side of the Infernal Regions." Among his exploits, Chamberlain learned to fight with an "Arkansas toothpick" (aka, a Bowie knife), with which he bloodied a comrade, and he fended off the amorous advances of a sixteen-year-old white girl (or so he said). Recalling these and other incidents, Chamberlain's book earned its subtitle, *Recollections of a Rogue*. Designed to shock, its scandalizing perspective would inflect subsequent cultural perceptions of San Antonio as a rough-and-tumble tourist mecca, licentious and libertine.[4]

Edward Everett, a military artist and draftsman serving with an Illinois regiment, was also stunned by what he encountered in the city. Unlike Chamberlain's gleeful response, however, Everett's account of its rowdiness was straight-laced and disapproving. Sent to watch over soldiers' behavior at local dance halls, his telegraph-like remarks set the disreputable scene: "The company attending these orgies consisted of Texans . . . teamsters, soldiers . . . without leave, Mexicans, gamblers and roughs. Of the feminine portion . . . I will say nothing; the perspiring throng . . . flavored with the fumes of whisky."[5] He drew greater comfort from another assignment, which would leave an influential impression of San Antonio. Everett's artistic skill caught the eye of a superior officer, who assigned him to create images that would convey the town's colonial history and its relics. In a series of sketches of the Alamo and Mission San José, and a set of others that later would illustrate George Hughes's report to the secretary of war, Everett produced some of the first detailed renderings of the Spanish-era missions that the American public would see. These buildings' crumbling character, especially the cannon-blasted façade of the Alamo, seemed to underscore the faded glory of a bygone empire. "The country bears evidence of having been at one period in a high state of cultivation and fertility," Hughes wrote in confirmation of Everett's artistic evocation, "supporting a large and concentrated population, who in time of danger sought refuge in the missions . . . monastic fortresses, whose stately and melancholy ruins attest their former magnificence and grandeur."[6] Future tourists complied after local Anglo promoters trafficked in this vision to rope in those very same visitors; in a slightly different way, so did well-intentioned preservationists who later fought hard to protect and rehabilitate the missions and their romantic appeal.[7]

A more pragmatic consequence of the war with Mexico was to convince the U.S. Army that the city had the possibility of becoming an important military depot. As General Wool marched toward the Rio Grande, Hughes and other engineers mapped the landscape and recorded latitude and longitude each day to orient those who trailed behind them. This topographical effort began in San Antonio, laying down the best routes through its southwestern hinterlands, a process that would pave the way for later railroad construction (and in the twentieth century, high-speed highways) along some of these same routes.

Hughes's reconnaissance affected the city in another way: to become a supply center it would need the resources to house, feed, clothe, and supply soldiers. With this in mind, Hughes scouted the entire community, noting which buildings "belonging to the government in the town might be conveniently converted into hospitals and barracks for a considerable force." The Alamo caught his eye, too: "if placed in a suitable state of repair, [it] would accommodate a regiment, and might at the same time be rendered a strong defensive work, well supplied with water."[8] The site would not become a fort, but the quartermaster's headquarters, and Edward Everett would be instrumental putting the building on a wartime footing. He developed plans for remodeling the Alamo's chapel and barracks and envisioned ordinance and medical storerooms, quarters for officers and men; shops for saddlers, carpenters, and blacksmiths; and sheds for forage, horses and stock. "Mexican" laborers would also be housed on site. Only some of these concepts were realized during the war—the barracks were shored up and made serviceable, while the debris-choked chapel was cleared out though not yet converted.[9] A more complete conversion would occur immediately after the conflict. An 1868 photograph indicates the new depot at work: A teamster had wheeled a covered wagon

Alamo and Wagons, c. 1868. *San Antonio Light Photograph Collection, UTSA Special Collections.*

up against the Alamo's front door, loading or off-loading material, and two other wagons occupy the foreground, waiting their turn. Above this scene, a large American flag flies, its flagpole jutting straight up from the army-commissioned curved parapet topping the original flat walls of the church's front face. That parapet, which has become an iconic symbol of the Battle of the Alamo, instead is a signal of the U.S. conquest of the Southwest in the post-Mexican War era.[10]

Local merchants took the hint, and Alamo Plaza and the surrounding streets began to fill up with vendors whose major business was with the U.S. Army: supplying feed, hardware, tobacco, and other essential services. Between 1848 and the outbreak of the Civil War, it was military disbursements that were largely responsible for the community's economic growth. A string of forts between San

Antonio and El Paso, designed to defend the newly conquered territory from Mexico and the Comanche, housed upwards of 1,500 troops (nearly 15 percent of the entire U.S. Army at the time); three years later Texas was home to 2,819 troops. To keep them supplied required a regular system of distribution along the new roads leading from the Gulf Coast to San Antonio, and then from the Alamo City west. The heavy wagons, laden with upwards of three thousand pounds of material, ground these new routes into the hard-packed soil. The distant outposts, and the transportation grid that linked them together, framed the "postwar military geography of Texas."[11] The costs incurred via this system of long-haul freight trains staggered Congress, but San Antonio reveled in the influx of cash. In 1853, for example, the army quartermaster reported spending "$110,000 in gold and silver per year on employment, rents, and purchases of goods and forage." Three years later, when it formally became the army's logistics hub for the western forts, the army employed 138 residents whose salaries totaled $39,000. At this time, too, the military contracted out its freight operations, injecting another stream of income into the city.[12]

So lucrative was the freight trade—civilian and military—that a battle broke out over its control. Tejanos may not have cornered the market, but they had been able to undercut most Anglo teamsters with their greater speed and lower costs. Amid the rise of the nativist Know-Nothing Party of the 1850s, and the continued racist anti-Mexican fervor that Anglos routinely expressed in Texas, in July 1857 white teamsters began to assault their business rivals. During the brief but bloody Cart War, Anglo-perpetuated violence included stealing freight, destroying carts, and murdering or maiming Tejano drivers. Tepid protests in local newspapers did little to stop the conflict. The Austin-based *Southern Intelligencer*, for instance, shared the generalized Anglo disdain for Mexicans but feared that this particular one-sided battle might foretell the outbreak of a more dangerous class struggle: "No generous mind could palliate this war upon a weak race, laboring as we all do for bread. If we permit the driving out of the Mexican laborers, a war upon the Germans will come next—and that must be followed by a war between the poor and the rich." It took a formal diplomatic protest from the Mexican ambassador to the United States, a plea from the U.S. Secretary of State to the governor of Texas to intervene, and the legislature's special expenditure for increased patrols to stop the hostilities.[13]

The damage had been done. Tejano carters largely had been driven out of this freight business, with two Anglos securing the military's contracts for hauling goods between the Gulf Coast and San Antonio and Austin. The sad result of the Cart War was just one way in which San Antonio's Tejanos were being eclipsed by newcomers. The new arrivals consisted partially of southern whites who had poured into the state following annexation and the conclusion of the U.S.-Mexican War, lured by the cheap price of land and who discovered Texas cities were rife with opportunities. German émigrés recognized this same main chance, with San Antonio becoming an epicenter of the waves of in-migration that began in the 1830s and accelerated across the 1840s. The so-called German Belt, a series of communities ringing the city, stretched from Castroville to the west to Fredericksburg, Kerrville, and Mason to the northwest, arching east to New Braunfels. Thousands settled in these farming communities, replicating the industrious agricultural life they had known at home. When Frederick Law Olmsted toured the region in 1853, he felt transported back to the Rhineland he had tramped through earlier in life. These new Texans produced large quantities of meat, hides, wool and other goods that they hauled to market in San Antonio. There they dealt with German merchants who had set up shop in the burgeoning city, sold grain to the German-controlled breweries and milling industries, and dined in restaurants and beer gardens that felt like the old country.

By 1850, Tejanos no longer constituted a majority of San Antonio's population (they would not regain this position until the second half of the twentieth century). While fighting to maintain their property rights, they battled against those who would deprive them of the right to the ballot (and won). When the anti-immigrant and anti-Catholic Know-Nothing Party captured city hall in 1854 and threatened to oust Tejanos from public life, José Antonio Navarro was among those fighting back. He led the drive to unseat the real Johnny-come-latelys, urging his community to recall their foundational role in the community's history: "Mexican-Texans are Catholic, and should be proud of the faith of their ancestors, and defend it inch by inch against such infamous aggressors."[14] The next year, Tejano votes powered the Democratic Party back into office.

Compared to the other cities in Texas, San Antonio of the 1850s was more ethnically diverse, a pluralism, in the words of religious scholar Timothy Matovina, that "showed a marked degree of separation between ethnic groups." These distinctions were "based not only on the different backgrounds of religious and ethnic groups, but also, in the case of Tejanos and Anglo Americans, on an underlying disagreement about which group constituted the dominant cultural force or legitimate 'host society.'"[15] Local Tejanos, "who did not cross a border to enter the United States but had the border cross them thought of themselves as the host society." Anglos believed otherwise: "once U.S. rule was established," they claimed "precedence as the harbingers of progress in the area."[16]

These lines of conflict and resistance became present as well in the city's increased spatial segregation. Tejanos dominated the western sector of town, in between the San Antonio River and San Pedro Creek, revolving around Main and Military Plazas and San Fernando cathedral. Germans began to cluster east of the river and then moved south of the central core—along König Wilhemstrasse (King William Street). Irish Flats lay east of the river and north of Alamo Plaza, while Anglos tended to live on streets radiating north. In this "jumble of races, costumes, languages and buildings," Frederick Law Olmsted spotted touristic gold. Its "religious ruins, holding to an antiquity, for us, indistinct enough to breed an unaccustomed solemnity; its remote, isolated, outpost situation, and the vague conviction that it is the first of a new class of conquered cities into whose decaying streets our rattling life is to be infused, combine with the heroic touches in its history to enliven and satisfy your traveler's curiosity."[17] In the coming decades Olmsted's observations, which confirmed and reinforced San Antonio's sharp distinctions by ethnicity, race, and class, would be repeated so often that they almost became a self-fulfilling prophesy, conceived by and written for those who had most to gain from it.[18]

The Civil War elevated these social and political tensions, though in San Antonio they played out in complex ways. The city did not have a robust connection to cotton production, which was the economic driver in the state's eastern counties; fewer than 600 slaves lived in Bexar County in 1860. The bulk of Texas's enslaved people were concentrated well east, and between 1845 and 1860 their numbers grew rapidly. In 1845, at the time of annexation, there was an

Photograph of a section of a lithograph titled *San Antonio de Bexar*, which reproduces various Hermann Lungkwitz paintings. The main illustration features Crockett Street looking west. *Ernst Wilhelm Raba Photograph Collection. Courtesy of the San Antonio Conservation Society Foundation.*

estimated 30,000 slaves in the new state; five years later more than 58,000 were clearing land and planting cotton and other crops. The 1860 census counted more than 182,000 individuals who lived in bondage, roughly 30 percent of the state's total population. That Anglo Americans enslaved so many offers disturbing confirmation of Stephen F. Austin's argument in 1824: "The principal product that will elevate us from poverty is Cotton, and we cannot do this without the help of slaves."[19] Slaves did not just work the fields. Their muscles and the crops they harvested gave white owners considerable economic and political power, power displayed in the growing sectional crisis that would erupt in civil war in 1861. As the anti-slavery movement grew in the northern tier of states, following the slow emancipation of slaves in the New England and Mid-Atlantic states in the wake of the American Revolution, many residents of these urban and industrializing regions began to attack slavery, the "peculiar institution." Southerners in turn crafted a pro-slavery rhetoric that defended their individual rights as masters and their regional economic interests. Tensions erupted across the 1850s as new states entered the Union: The Compromise of 1850, like "Bleeding Kansas," where (ironically) the compromise and the subsequent Kansas-Nebraska Act (1854) helped to bring on bloodshed later in the decade, became flashpoints, as did the Dred Scott decision (1857) and John Brown's assault on the Harper's Ferry Armory and Arsenal (1859). Political parties imploded, new factions, including the Republican Party, coalesced around sectional concerns—a splintering of institutions and values to such a degree by 1858 that New York Senator William Seward proclaimed that the United States faced an "irrepressible conflict."

Texas was similarly torn. The eastern and Gulf Coast counties were hotbeds of pro-southern agitation, but frontier counties contained a significant amount of Unionist sympathizers. These distinctions became important as the debate over secession exploded in the winter of 1860 following Republican Abraham Lincoln's election as president in that November. Within a month, South Carolina had voted to secede, with Georgia and the Gulf Coast states—Florida, Alabama, Mississippi, and Louisiana—doing so in January 1861. Texas Governor Sam Houston tried to short-circuit the secessionists' demands for a statewide convention by calling the state legislature into special session in hopes that it would delay the process. Dominated by the more populous and slave-owning eastern portions of the state, the representatives called for a convention in late January with a subsequent popular referendum on its resolutions. The convention resolved overwhelmingly to secede, and on February 23, 1861, the voters across the state concurred by nearly a four-to-one margin.[20]

The margin was much narrower in Bexar County, with 827 voting for secession and 709 opposed. Some Tejanos were in support (and many of them would serve in the Confederate Army), while others were not.[21] Anglos from the Deep South were practically unanimously pro-secession, but most who migrated from the North and many from Upper South states like Kentucky and Tennessee remained committed Unionists (and some, like financier George Washington Brackenridge, would leave the city for the duration of the war). The Germans also tended to remain loyal to the Federal cause, though not exclusively so. The day before the formal secession vote, the Unionists put on a public display of their convictions, massing for the "largest procession of citizens ever witnessed in the town, with and without arms, [to have] paraded through the principal streets."[22] The next day's electoral result, however close, was also contested. Some Unionists were convinced that the proslavery faction, particularly members of the local chapter of the Knights of the Golden Circle (KGC), a secret society that had been agitating for secession since its founding in 1854, had committed voter fraud.[23] "Mexican votes [were] purchased at the rate of two shillings a piece," one disgruntled Unionist wrote to the *New York Times*, and there were "all kinds of irregularities and frauds practiced by the officers at the polls, all of whom were K.G.C.'s."[24]

Whatever the truth of these allegations, it did not alter the results. Even before the election Confederate forces had taken control of the Alamo quartermaster depot and its supplies as well as the U.S. Army's arsenal with all its guns and artillery, ammunition, powder, and shot. Despite the new flag that flew over the Alamo—on February 18 "the Stars and Stripes upon the old Alamo were torn down" and in its place "the Lone Star flag was hoisted"—ani-

Surrender of the Alamo to secessionist troops, February 16, 1861. *Harper's Weekly, March 23, 1861.*

mosities continued. Unionists were harassed in the streets, harassment that escalated following the April attack on Fort Sumter in Charleston, South Carolina, which ignited the Civil War. Jean-Charles Houzeau, a Belgian scientist and abolitionist, likened the violence in San Antonio to a reign of "White Terror."[25] He joined a small underground movement that ferried fugitive slaves and white Unionists to Mexico, fleeing with them when he faced imminent arrest.[26] The Germans bore the brunt of these threats in San Antonio and out in the Hill Country villages they had established in the preceding decade. The region's one significant battle—known alternatively as the Battle or Massacre on the Nueces—involved several dozen German Texans and a few Anglos who had decided to flee to Mexico. Hoping to capture them, a Confederate unit, attacking in the dead of night, surprised the Germans. Most of the Germans escaped, but they left behind nineteen killed and wounded. The victors then executed each wounded captive with a single shot to the back of their heads, leaving their bodies where they had fallen. "This was no battle," one analysis has concluded. "It was murder. It was a war crime." In the following weeks Confederate forces hunted down those believed to have been involved with the Unionist resistance, imprisoning several and executing others.[27]

The brutal turn of events did not shock the *San Antonio Express*. The day before the first Confederate cannon fired on Fort Sumter, it had castigated the Confederacy as "conceived in sin, shapen in iniquity, and born out of due time, because it was rushed into the world with indecent haste expressly to prevent people from beholding its deformities."[28] The newspaper's pages also record a community that made its peace with the war. An ammunition factory churned out cartridges for rebel rifles, a tannery produced enough leather for the annual production of 180,000 pairs of shoes, and its merchants made bank, an incipient industrialization that gained steam after the hostilities ceased. Far from the major battlefields, and with well-used roads to the Rio Grande Valley, local commercial interests and freighters set up a system of transport that funneled cotton from Arkansas, Louisiana, and Texas to Matamoros, Mexico. From that bustling Mexican port cotton-packed ships sailed to Liverpool, England, evading the Union Navy's blockade of Gulf ports. Matamoros also served as the transshipment node for badly needed supplies that San Antonio's wagons and carts carried back north for sale and further distribution. South Texas ranchers contributed to the local economy as well, running herds through San Antonio to feed Confederate troops in East Texas and Louisiana, a trade that expanded rapidly after the war.[29] As in the past, so in the future, military conflict had proved a boon to San Antonio's economic fortunes, a munificent reality that impelled most of its residents to cheer the bugle-blaring return of the victorious Union Army on August 2, 1865. One of the vanquished, who had served as a Confederate ordinance officer at the Alamo, and who watched Old Glory once again rise above the old mission, was William Giles Martin Samuel.

5
Recovery and Development

The rains came slashing down during a powerful late March 1865 storm, and the runoff slammed into the upper reaches of the San Antonio River and San Pedro Creek, sweeping up downed trees and other debris. As this churning surge powered downstream, it rolled over the river's banks in part because some of the detritus, "a great mass of driftwood forming a dam and backing up the water," piled up against the Houston Street Bridge's midstream support structure. The bridge held, but the floodwaters killed three people, submerged streets and alleys, devastated homes and buildings, and spread sewage everywhere. It took days for the fetid waters to recede. By then, the community was agitated. Those with long memories recalled the devastating 1819 flood that had destroyed the then-Spanish town; more recently, in 1845 and 1852, other floods had inundated the city. By 1865, many of the town's nearly 10,000 residents no longer could accept this pattern of destruction as a condition of living in a floodplain. They demanded action, calling for a dam to span Olmos Creek north of its confluence with the San Antonio River. Theirs was a smart proposal and local officials hired three engineers to assess the likelihood that such a flood-control structure would achieve the desired end and do so in a cost-effective manner. Their final proposal was replete with suggestions to clear away obstructions—natural and human—in the river's and creek's channels and to build bridges that did not require debris-blocking supports. These small fixes would be far cheaper than an upstream dam, the report indicated, but even so the city enacted few of the suggested changes. It was not until the 1920s, after another killer flood blasted through the city, that a galvanized San Antonio finally put up the money to construct the Olmos Dam.[1]

STREET SCENE IN SAN ANTONIO.

From the Edwards Aquifer Website: "This engraving entitled *Street Scene in San Antonio* helped illustrate an article called 'Through Texas' by Frank H. Taylor in the October 1879 edition of *Harper's New Monthly Magazine*. The depiction of the river as a 'street' points up the reliable stream's importance for 19th-century commerce and transportation in San Antonio"; <http://www.edwardsaquifer.net/sariver.html> [Accessed Apr. 3, 2018]. *Courtesy of the Edwards Aquifer Website by Gregg Eckhardt (http://www.edwardsaquifer.net/).*

This incident could serve as a tagline for the ensuing decades: The citizenry had a well-defined sense for what it needed to transform the community, to make it more salubrious, enhance its economy, and increase its efficiency and effectiveness. It just had a difficult time agreeing on the means to these laudable ends.

Consider the railroad, which finally arrived in 1877. With the exception of those who made their living using horses, oxen, and mules to haul freight to and from the Gulf Coast or down to the Rio Grande Valley, every merchant, commercial agent, and entrepreneur recognized that the iron horse was the engine of nineteenth-century progress. All they had to do was see how their in-state rival,

Houston, had profited from the construction of a series of pre-Civil War rail lines that linked its hinterland to that city's intensifying urban marketplace. In 1852, the City of San Antonio and Bexar County each invested $50,000 in the San Antonio and Mexican Gulf Railroad, the brainchild of a group of local entrepreneurs. Its projected route would run southeast to Victoria and then on to either Indianola or Port Lavaca on the Gulf, but the project came to naught. Victoria and its eponymous county refused to pitch in until rails reached the Guadalupe River, and the two port cities fought over which one should be the terminus. As eager as San Antonio's capitalists appeared to be about the railroad's prospects, they did not have enough money to build the road on their own. "Instead of the promised twenty miles of track each year," historian Donald Everett observed, "the only apparent activity was just outside the city where workmen cut some underbrush and ploughed a furrow to comply with legal technicalities." For the next twenty-five years and more, the city and county were embroiled in a lawsuit with the railroad corporation, an entity that in 1867 the state courts had determined was "rotten from the beginning."[2]

Other initiatives came up short because of the Civil War and the Panic of 1873. By the mid-1870s, a new effort that northern capitalists had reorganized from preexisting lines—the Galveston, Harrisburg, and San Antonio Railroad—began to lay down tracks. Before committing to San Antonio, the owners required a substantial local commitment of money from Bexar County, which resulted in a sharply contested bond election in January 1876. Although the railroad's supporters won a narrow victory, the county would struggle to sell the requisite $300,000 worth of bonds. A year later, as the construction neared town, there was a palpable sense of celebration and foreboding. "Already does the shrill whistle of the locomotive reach our ear," the *San Antonio Express* reported, and it "speaks of an enterprise whose . . . spirit of progress has begun to arouse the old, dry bones of the Alamo City." Yet while it cheered this new energy, the newspaper also sounded a cautionary note: "The era of prosperity we have for the last score of years labored to bring about is achieved, though as yet in its merest incipiency. What remains to be done, is incumbent upon the people." The people were, as experience had shown, a fickle lot.[3]

Their political engagement was just as capricious, or at least that was true of those for whom politics was the game of choice. Local white Republicans, a small but feisty collection of activists, were among those playing by the new rules emerging during the postwar era. In 1867, the state's military governor dismissed the elected mayor and Board of Alderman (who represented the city's four wards) and appointed Republicans in their stead. Maintaining their power was complicated. Republican commitments to protect the newly enfranchised African Americans, whose votes they assumed would keep them in office, ran into a set of intra-party squabbles and external pressures that narrowed their options. To navigate this fraught political landscape, as James Pearson Newcomb discovered, was tricky. An anti-secessionist before the war, and after it a Radical Republican, Newcomb quickly modified his initial conviction that black empowerment was right and good. To gain the support of the German American community in San Antonio—one of, if not the, largest voting bloc in the city—Newcomb had to convince them (and other whites) that Republicans could "control" African Americans. Through the Loyal Union League, he and other white Republicans registered black voters with the objective of harnessing them to Republican objectives. As wards of the party, they would be treated, the *Express* assured its readers, "the way society does orphans."[4]

However paternalistic this language, it did not prevent Republicans from being smeared by Democratic race baiting. Former mayor Samuel Maverick, for one, argued that Newcomb and his white allies were "in league with that arch traitor from hell Thaddeus Stevens," a Radical Republican in the U.S. House, a fierce opponent of discrimination and a staunch advocate of the freedmen's rights. Maverick was among those calling for the potential use of force in defending white "civilization" from black rule, and in the spring of 1868, a chapter of the Ku Klux Klan emerged. Its appearance won the praises of the *San Antonio Herald*, the Democratic mouthpiece: "If the Ku Klux's have for their objective the protection of the men, women and children of the white race against the brutality, rapacity, and ferocity of the black barbarians, we are with them, of them, and for them, first, last, and all the time."[5]

This verbal violence found physical expression on the streets whenever black voters went to the polls in the late 1860s; along the way, they were harassed and beaten. Worse, they gained little support from the very Republicans who were desperate for their votes. Scoffed the *Express*: "The rebels represented by the *Herald* of this city . . . are becoming alarmed at the prospect of 'niger rule' and cry out miscegenation," an unfounded worry, as Republicans had no more interest in racial mixing than their opponents. As a result of Republicans' deeply felt paternalism and political calculations in combination with German American skepticism of radical social experiments and questions of political equality, Newcomb and others shelved earlier notions of promoting a more equitable San Antonio. In so doing, they essentially gave the keys to the city to German American politicians, who dominated city hall for the next twenty years.

That compromise undercut another Republican aspiration, to build a more energetic and diverse economy. The "frugal and sparing" German contingent presented a united front in favor of lower taxes and fewer services and in opposition to investments in railroads, streetcars, and sidewalks.[6] Except, that is, for those urban development schemes that benefitted their interests. Because German "businessmen rapidly turned Commerce Street, the city's major commercial artery, into their private preserve," writes historian David Johnson, the city council gladly underwrote improvements to the street and compensated property owners for losses, real or potential, due to street repairs.[7] They moved rapidly when another of their key interests was threatened. By 1868, the U.S. Army had outgrown its cramped quarters at the Alamo and surrounding buildings and asked the city for a larger, stand-alone base outside of town. The request came with a warning: if the city failed to donate land for a new facility, the army would move the depot to Austin. Given the close connections between the German merchants, their fellow émigrés farming and ranching in the nearby Hill Country, and the U.S. Army's growing need for forage and feed, horses, mules, wagons, uniforms, and other critical supplies, the council responded positively. It agreed to set aside more than ninety acres for a new military reservation on what locals would call Government Hill, high ground overlooking San Antonio to the east of downtown. By the mid-1870s, Congress had appropriated $150,000 for the construction of the new army depot, commodious stables and parade grounds, another $125,000 built fifteen limestone houses for officers, and a third infusion of federal funds in

1883 purchased more land and the construction of a base hospital. As would become true of later investments that advanced San Antonio's tight relationship with the nation's armed forces, the *San Antonio Express* cheered this immediate outcome: "The building of these barracks and quarters will enliven businesses on government hill considerably and create a new boom there both in building and real estate."[8]

This particularistic approach to governance and growth privileged some San Antonians over others. The disadvantaged included those already marginalized. African Americans, who by 1880 accounted for 14.8 percent of the city's population, resided in each of the four wards, where they rented shacks, stables, or garages that fronted back alleys. Living "in clusters of families, relatives, and friends residing among and near their white employers," and occupying substandard dwellings, offered them ample "physical and visible evidence to reinforce the mostly negative stereotypes" they endured.[9] Segregation intensified in the last decades of the nineteenth century, as African Americans increasingly concentrated on the city's east side, in a neighborhood known as the Baptist Settlement—so named for the black congregations that were located there. These religious institutions provided a much-needed sense of community in a semi-private matter, while the annual Emancipation (Juneteenth) Parade, which worked its way from Main and Military Plazas to San Pedro Park, gave public expression to the presence of the city's black community. Regardless of whether they worshipped on their home ground or marched through civic space, the city's black population remained acutely aware of their second-class status, of "the restrictive nature of urban life."[10] The Rev. Mark Henson of St. Paul's Methodist Church made the same case in his 1884 speech in San Pedro Park at the conclusion of that year's Emancipation Parade. "It is [the black man's] lot to live in a land where all presumptions are arrayed against him, unless we accept the presumption of inferiority and worthlessness. If his course is downward, we meet little resistance," Henson declared. "But if upward, his way is disputed at every turn in the road . . . he excites resentment and calls forth stern and bitter opposition. If he offers himself to a builder as a mechanic, to a client as a lawyer, to a patient as a physician, to a university as a professor, or to a department as a clerk, no matter what may be his ability or attainments, there is a presumption, based on his color or previous condition, of incompetency, and if he succeeds at all, he has to do so against this most discouraging presumption."[11]

The Tejano community met with many of the same constrictions. The 1880 U.S. census confirmed that it had fallen to the third largest ethnic group in San Antonio, and that its members had squeezed into a single ward covering the city's west side. They congregated around parish churches and were active in religious and social organizations that enabled them to forge close kinship and communal ties, celebrate feast days, dance the fandango, and honor the Virgin of Guadalupe. But most lived on unpaved streets, lacked readily accessible potable water, and shared their streetscapes with innumerable bars, honkey-tonks, and saloons. Theirs was also pestilential terrain, with local health officials reporting spikes in cases of dysentery and tuberculosis on the west side that politicians routinely ignored, a stark reminder of how rapidly Tejanos had lost political ground since 1845.[12]

The tourist industry reified their diminished status when it slapped a patina of romance over their impoverished condition. No sooner had the first train arrived in 1877 than assorted journalists and travelers began sending reports east about the Alamo City's exoticism. In Military Plaza, the open-aired marketplace, they ate chili that "little black-eyed girls" served, and whose "every movement is jealously watched by beetle-browed hags squatting beside the fires, or by dark-visaged men who lounge in the near-by shadows." They attended feast days and fiestas, cockfights and fandangos, especially marveling at the latter: "Mexicans of all classes seem to me to be infatuated on the subject of dancing," one visitor wrote. "Perhaps there is no other people in the world of whom it may be so truly said that their genius lies in their heels."[13] San Antonio's booming trade in visitors "depended upon images of Mexican culture to market the city visitors even as Mexicanos were themselves marginalized."[14]

This touristic appropriation of "old" San Antonio—its mysteries and marvels—formed the backdrop for a second appropriation known as the Battle of Flowers parade. The brainchild of elite white women, who inaugurated the parade in 1891, it commemorated the Battle of San Jacinto, which had been fought fifty-five years earlier. In the process, the annual parade enshrined the Texas Revolution as

Photograph of the chili stands on Military Plaza, looking north. *Ernst Wilhelm Raba Photograph Collection. Courtesy of the San Antonio Conservation Society Foundation.*

a marker of the subsequent Anglo dominance of San Antonio. The parade's original route, with flower-draped wagons and carriages rolling out of Alamo Plaza and wending their way to Main Plaza and San Fernando Cathedral, then circling back to the Alamo, reinforced this claim. In its movement through the city's segregated geography, the elite attempted to naturalize their position of power: "they connected the certainty of historic commemoration with the continuation of their own roles as cultural guardians." At the same time, these self-appointed keepers of the flame underscored their paternalistic guardianship as they pushed to reclaim and preserve relics of the Spanish past.[15]

That modernity required racial subordination and ethnic marginalization was a notion that would not have surprised white visitors or those locals who profited from their presence. San Antonio required Yankee ingenuity and enterprise, wrote Bostonian Harriet Spofford, whose husband was the lead investor in the Galveston, Harrisburg, and San Antonio Railroad. It needed "something as powerful as a norther to push through all the sinuosities of her countless streets."[16] It was northern consumers whose rapacious demand for Texas beef drove the legendary cattle drives from South Texas to San Antonio (where free-spending cowboys and vaqueros spent considerable money in the city's mercantiles and its renowned dens of iniquity crowding the west side), before following the Chisolm Trail north to railheads in Kansas.[17] It was northern capital that underwrote the first, and then second, third, and fourth railroad lines that by 1890 tied San Antonio into the national economy.[18] It was outside capital that funded the city's emerging streetcar network that linked the railroad depots to major commercial sectors and neighborhoods; its needs compelled the straightening and widening of central arteries, a reconstruction that writer Stephen Crane bemoaned. He arrived in January 1889 on a quest to immerse himself in "the poetry of life in Texas," yet found to his dismay that San Antonio had become an earnest American city. The "serene Anglo-Saxon strings telegraph wires across their sky of hope; and over the energy, the efforts, the accomplishments of these pious fathers of the early church passes the wheel, the hoof, the heel." Ruing the "almighty trolley car," he denounced these "merciless animals" and their capacity to "gorge themselves with relics. They make really coherent history look like an omelet."[19]

One piece of that heralded Spanish past, the city's water system, required updating. The acequias, irrigation ditches that the mission Indians had constructed in the early eighteenth century, had remained the city's central means of distributing water for domestic and agricultural purposes. Although brilliantly conceived and engineered, the system could not meet the needs of the growing nineteenth-century city. By the 1840s, the acequias' limitations were beginning to be recognized. One of these limitations is that residents used the ditches to draw in water and flush out waste, a practice that compromised public health and which officials believed was responsible for the spread of cholera. Epidemics in 1846 and 1849 made it clear that a more sanitary system was essential if the San Antonio was to escape another fatal encounter with the water-borne disease. Nothing happened. A third major epidemic in 1866, which lasted for two months and killed 198 residents, led to renewed calls for action. "Conditions so long deplored by social reformers still existed: muddy streets, filth-infested vacant lots, polluted river water, and grossly ineffectual methods of sanitation." The city did not act for more than a decade, a result of political squabbling, lack of taxpayer support, and a crimped local treasury. When at last, in 1877, a group of private investors received a contract to develop a modern system, complete with pumps, raceways, and reservoirs, it appeared as if the long-awaited fix was at hand.[20]

Although George W. Brackenridge did not start the San Antonio Water Works Company—that honor lay with a French immigrant, Jean Baptist Laconte—he quickly became its majority stockholder and president. His relationship to the city's major river, and thus its main source of water, was not just financial, but personal. His family owned the property that surrounded the river's headwaters and, as such, he controlled its flow. This rankled city hall, but an earlier generation of officials had off-loaded that land, and in time, Brackenridge's mother purchased it. Thus, when Laconte's company secured the contract to provide "pure" water to the community, the only clean water readily available was that which headed downstream from the Brackenridge's property. Complications ensued, as they always did in cash-strapped, faction-ridden San Antonio, but the upshot was that by 1886, the new waterworks was under Brackenridge's ownership. He expanded the system to include two pump houses and reservoirs, and the local politicos hated him (even as the city kept going to him for banks loans to bail it out of its financial difficulties). Denouncing Brackenridge as a "monopolizer" (he owned the San Antonio National Bank, the *Express*, was a major landowner, and a rival newspaper estimated his wealth that year at more than one million dollars), Alderman Nelson Mackey promised to fight Brackenridge tooth-and-nail: "If Brackenridge owns the head of the river, he can govern the city by curtailing the supply of pure water. If the city owns it, we can govern him."[21]

That tough talk never panned out, and in 1893 Brackenridge added to his water dominance after drilling a deep well downtown that struck white gold. "The fresh water supply of San Antonio is apparently unlimited," enthused the *Express*. "It has increased three million gallons for each twenty four hours by a splendid strike in the artesian well being drilled on the property of Colonel

San Pedro Springs, 1877. *Courtesy of the Edwards Aquifer Website by Gregg Eckhardt (http://www.edwardsaquifer.net/).*

George Brackenridge on Market Street," a productivity that "was one of the most complete successes and marvels yet discovered in San Antonio."[22] By 1900, the city of 53,321 no longer depended on the river for its potable water and the acequias were officially decommissioned. That said, there was a serious environmental cost to the extensive pumping of the aquifer, which emerged as other water prospectors sank one well shaft after another into the artesian plain. Local springs and seeps started to disappear; the levels of the San Antonio River and San Pedro Creek, whose streamflow arose out of the Edwards Aquifer that these new wells were tapping, declined. Brackenridge understood what was happening, and wrote a letter of lamentation to a close friend: "I have seen this bold, bubbling, laughing river dwindle and fade away. It is now only a little rivulet, whose flow a fern leaf could stop and its waters are hardly enough to quench the thirst of a red bird. This river is my child, and it is dying, and I cannot stay here to see its last gasps."[23] He did not depart, but would unload some of his assets: Brackenridge sold his waterworks shares to other investors, donated hundreds of riverfront acres to create a city park (later named for him), and sold property to the park's north to the Sisters of Charity of the Incarnate Word that would become the University of the Incarnate Word. Meanwhile, it was not until the 1920s that the city finally bought out the private water company and turned it into the City Water Board. Even this gesture toward public ownership of a key utility faltered in its civic duty. For the next forty years, the Water Board delayed underwriting of a citywide water-distribution system.[24]

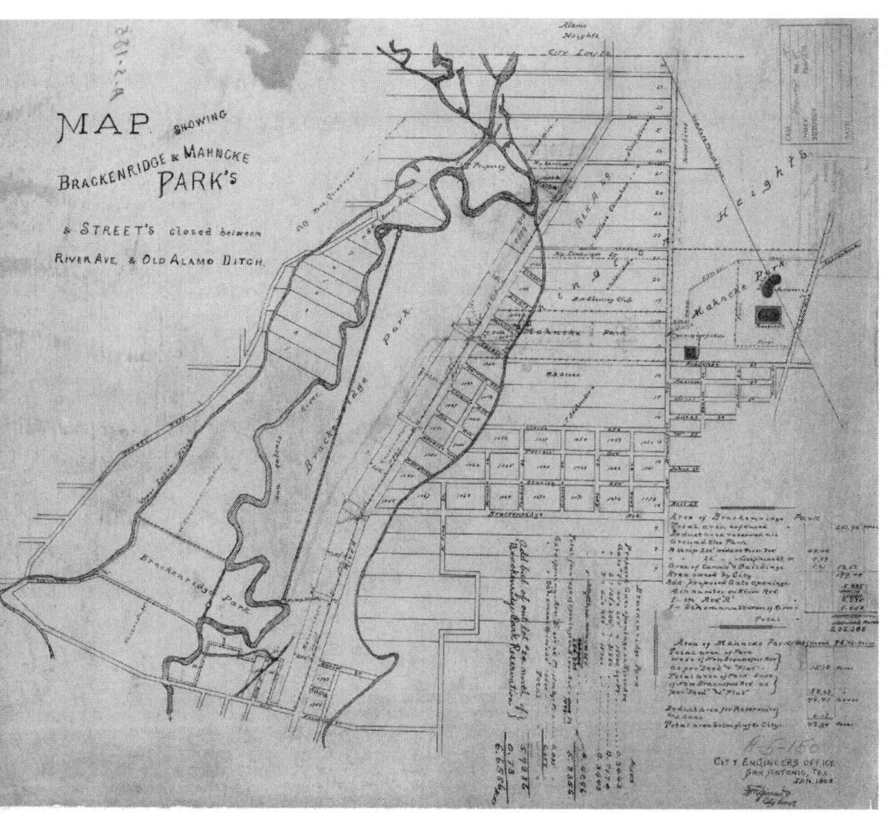

1908 map of Brackenridge Park. *Courtesy of the Edwards Aquifer Website by Gregg Eckhardt (http://www.edwardsaquifer.net/).*

That public investment did not match—and did not try to match—private investment in the city was because few local powerbrokers conceived of their interest as anything larger than personal. The city may have "welcomed growth, but [it] did nothing to promote or to provide for it," argues historian David Johnson, a chilling indication of San Antonio's divided character and divisive politics.[25] Bryan Callaghan Jr. hoped to change the city's crippling insularity, and he had the pedigree to do so. His father, Bryan Callaghan Sr. was a successful merchant and former mayor; his mother, Concepción Ramón, was a member of the Tejano elite. A Catholic and a Democrat, Callaghan was reputed to have been fluent in the city's four main languages—English, Spanish, German, and French. His close ties to the west side working class gave him his start in local public affairs, winning election as their alderman in 1879. Six years later, he won his first term as mayor and embarked on a significant program of developmental projects. Between 1885 and 1893, the mayor granted franchises for streetcars, gas lines, and lighting, sold bonds to pay for the construction of a new city hall dropped in the middle of Military Plaza, and took out loans from local banks to pay for improvements rather than raising taxes. Like Alderman Mackey, he, too, routinely railed against George Brackenridge; the banker countered that Callaghan and his ilk were happy to receive loans; they just hated paying them back.[26]

When Callaghan resigned as mayor in 1893 to become county judge, the chief administrative officer of county government, he had done much good and run afoul of a new set of particular interests. Those most particular about the need for change were the swelling population of Anglos who as of the 1890 census constituted the majority of the city. This ethnic bloc of voters, many of whom favored prohibition, disdained Callaghan's multi-ethnic, multi-ward coalition, and the political support it received from the hard-drinking town. They had a point: in 1883 the *San Antonio Express* observed that while all saloons would be closed on election day, they would be "Very likely as usual, open at the back door. As a general thing, there are more drunks on election day than any other."[27] Unable to swallow that hypocrisy, temperance advocates contributed to a new political reality that pitted "Baptists against Catholics, Anglos against Germans and Mexican-Americans, pro-

George W. Brackenridge. *Prints and Photographs Collection, di_02398, The Dolph Briscoe Center for American History, The University of Texas at Austin.*

viding a basis for community factionalism that Callaghan ultimately could not control."[28]

These brawls, with Callaghan at their center, extended into the twentieth century and would lead to a rewriting of the city's charter and the emergence of a new kind of politics. Such contentiousness was not what caught the eye of a local reporter, however, who was eager to report on the progress that defined San Antonio as it approached the new century. The construction of new schools and churches meant a more educated (and presumably) moral society. Gleaming buildings rose above once-neglected thoroughfares, fire hydrants were posted along a streetscape lit at night by gas-flaring lamps. Macadam now covered downtown roadbeds, curbs rose up to hardened sidewalks: in the "'good old days,'" he reminded his readers, only the plazas had sidewalks. "They were fearfully and wonderfully made of flagstones carefully balanced on an unseen rock and with just that amount of tilting power that would enable them in wet weather to fill the pedestrian's pant legs with muddy fluid and his mouth with double-distilled profanity." Even flooding concerns had receded. Once the "great bugbear of the inhabitants of San Antonio," leading those who remembered the great flood of 1865 to gaze anxiously out their office windows and then "figure out on their maps the certain destruction which was to come," those anxieties seemed a thing of the past, the future seemed more benign. Although the city had not spent a penny on upstream flood control, the self-satisfied journalist opined, "every possible danger from the Olmos has been averted merely by converting the hard, unbroken prairie into cultivated fields. The man who would venture to predict an overflow today would be called an idiot."[29]

6
A New Day

It had been decades since armed men in large numbers had entered San Antonio. Many of the city's older residents in 1898, observing the First U.S. Volunteer Cavalry detrain that May, would have recalled the violence that had erupted in and around the Alamo City during the Civil War. Fewer were alive with direct memory of two moments in the 1840s when gunfire erupted: the Council House fight of 1840 and the Mexican army's reinvasions of the Republic of Texas in 1842. By contrast, the recruits who arrived in the late spring of 1898 came in peace. At first glance, though, the 1,250 men who had mobilized to fight in the Spanish-American War looked as if they had turned the page back to the city's desperado past. Many hailed from ranches and small western towns, had been sheriffs, deputies, and cowboys, and they were itching to go to war: "rough, tough, we're the stuff," they chanted. "We want to fight and we can't get enough." In their zeal, they initially called their regiment "Teddy's Texas Tarantulas," before settling on "Roosevelt's Rough Riders." Their commanding officer, Lt. Colonel Theodore Roosevelt, shared their bravado. As a child, he later declared, he had "thrilled and tingled as I read of the mighty deeds of Houston, of Bowie, of Crockett, of Travis, of the men who were victorious at the fight at San Jacinto, of the even more glorious men who fell in the fight of the Alamo."[1] Roosevelt had been so eager to engage in the war against Spain that he had resigned as President McKinley's assistant secretary of the navy, despite the commander in chief's pleas that he remain at his important post. "Theodore is wild to fight and hack and hew," a friend observed. "It is really sad. Of course this ends his political career for good."[2]

Theodore Roosevelt posing on horseback, in uniform, between two other men on horseback, in San Antonio during the Spanish-American War, 1898. *Library of Congress Prints and Photographs Division, Washington, D.C.*

Roosevelt was convinced that the reverse would be true. He was right: on his heralded return from battle, he promptly won election as New York's governor, then became McKinley's vice president, and, after an assassin's bullet cut down the president, ascended to the White House. That possible future did not distract Roosevelt from the task at hand when, on May 16, 1898, he and his men made their way down to Riverside Park to set up their training grounds. The 350-acre park, like most other open space that the city had acquired in the nineteenth century, had been a gift, in this case an 1887 donation from the San Antonio and Aransas Pass Railroad. Its "stately pecan trees, shady groves, and picturesque river," the *San Antonio Express* cheered, "can make it as delightful a retreat from the heat and dust of the city, as one could wish to see."[3] That was not how Roosevelt and the Rough Riders experienced the site. For two weeks, the "sun blazed down" and thick swirls of dust engulfed the perspiring men: "At supper time so much dust would be showered over the mess tables to such an extent that the men would sometimes have to eat their rations inside the barracks."[4] San Antonio proved a grueling environment in which to drill, test weapons, and practice scouting and skirmishing. There were nightly compensations: the city opened its doors to those some later would proclaim the heroes of San Juan Hill. Roosevelt reveled in the admiring attention he and his sweat-stained men received on the streets and in the bars, a genial interlude before going into battle.[5]

Theirs was a one-sided fight, and its consequences were massive. The Spanish-American War turned the United States' longstanding international aspirations into a done deal. Roosevelt hungered for just such imperial ends, and touted its unsavory sidekick, the White Man's Burden. As the Rough Riders rolled out of San Antonio, on their way to additional training in muggy Tampa, Florida, from which they would embark for Cuba, their commander lectured them on the benefits of social Darwinism. The natural world offered profound lessons for the human. In each, the "survival of the fittest prevails," he asserted. "Look at the magnificent trees along the river," he noted as the train chugged across the Sabine River in East Texas. "The ones that started out crooked were crowded out and died. The strong and straight saplings appropriated all the food." The same held true in international affairs: some nations were stronger than others, their moral fiber and social virtue were more developed than others. The United States was fit; Spain was not. Roosevelt implied that Spain would suffer at war's end in ways similar to what happened among "wild animals"—"the cripples and inefficients that cannot support themselves are killed off."[6] The cost a defeated Spain actually would pay was touted up in the number of former colonies it lost via the Treaty of Paris (1898). For a lump sum of $20 million, Spain relinquished its hold over the Philippines and Guam in the Pacific as well as Puerto Rico, Cuba, and its possessions in the West Indies. The United States had become a world power.

San Antonio, which early in the nineteenth century had been a Spanish outpost on its North American frontier, at century's close had become an agent in the disintegration of what remained of Spain's colonial empire. Its contributions to this particular war's outcome would be replicated in its enduring support of the nation's now more extensive global interests in the twentieth century. President Theodore Roosevelt, who returned to San Antonio in 1905, cheered these new international engagements in a speech before the Alamo, asserting that they were important extensions of the Monroe Doctrine. "The only thing we can decide is whether, being great, we will do well or do ill," he thundered. "We must so handle ourselves that no weak power which is behaving itself shall have cause to fear us; and no strong power of any kind shall be able to oppress or wrong us. We have duties in connection with the great position we have taken. We cannot shirk these duties."[7]

His tour of an expanding Fort Sam Houston (known colloquially as "Fort Sam") underscored how significant this base had become to the nation's international commitments, its capacity to project power forward. It served as a training facility for thousands of troops that the United States had stationed in the Philippines and in the Caribbean to defend its possessions, pumping more than $1 million into the local economy.[8] In 1905, it became home to new cavalry and light artillery missions; the next year it created a satellite firing range and training ground northwest of the city, later named Camp Bullis. The local U.S. Army arsenal, sprawling over twenty-one acres just south of downtown, produced munitions for American forces battling ongoing revolts in the Philippines and elsewhere. (At its peak during World War II, the arsenal produced an estimated 337,000,000 pounds of ammunition).[9] These streams of men and material widened each time President Roosevelt, and his immediate successors, William Howard Taft and Woodrow Wilson, dispatched combat-ready forces to back up the nation's imperial mission. Between 1900 and 1917, the United States stationed troops in Puerto Rico, the Panama Canal Zone, and Cuba; established formal protectorates over Haiti, the Dominican Republic, and Panama; and periodically occupied each of these countries as well as Nicaragua, Honduras, and Mexico. A large contingent of these soldiers first earned their stripes in Fort Sam's classrooms, at its firing ranges, and on its parade grounds. When in 1916 President Wilson ordered the U.S. Punitive Expedition to suppress Mexican incursions that threatened U.S. border towns, these

Quadrangle and Staff Post, Fort Sam Houston, ca. 1893. *General Photograph Collection, UTSA Special Collections.*

men sallied forth from Fort Sam, which was then the command headquarters for the army's Southern Department and the largest post in the United States.

For all the turmoil and possibilities that the Spanish-American War had let loose locally, growing the city's economic fortunes ultimately depended on hometown politics and entrepreneurial initiative. Therein lay the rub. Without much effort on the part of the city, San Antonio secured government contracts in support of the military's operations and logistical needs, during both war and peace. For some business leaders, it made little sense to develop a more proactive approach to growth and development, particularly if that led to increased expenditures and taxes. This is one of the reasons why the city, in 1900 the largest in Texas, lacked the industrial base and capital resources that Houston and Dallas enjoyed.

It was also one of the underlying tensions that continually roiled local politics during Bryan Callaghan's first stints as mayor. Even as his opponents attacked his use of patronage to secure electoral victories—he was not called Boss Callaghan for nothing—they also assailed his willingness to incur debt to finance civic improvements. Inspired perhaps by the increasing chorus against his tactics and objectives, Callaghan's 1893 jump to county government was only temporary. Four years later, he was reelected as the city's mayor only to lose the post in 1899, after a broad coalition successfully attacked what they decried as his politics of cronyism. Of greatest concern to this generation of self-appointed reformers was the political system, and ethnic voters, that kept Callaghan in office. His commitment "to decentralized, ward-based politics" clashed with their progressive leanings that called for a greater centralization of power, much as their hero, Theodore Roosevelt, advocated nationally.[10]

The political system that seemed to provide the consolidation they desired was the commission form of government, which had been developed in the tragic wake of the massive hurricane that destroyed Galveston, Texas, in 1900. The storm killed more than six thousand people, obliterated the built landscape, and crushed the city's economy, the worst such disaster in U.S. history.[11] As part of the rebuilding process, surviving members of the city's social and economic elite conceived of a new form of governance—often referred to as the Galveston Plan—that would consist of five commissioners the governor would appoint, each of whom would oversee specific aspects of municipal operations: public safety (police and fire), public works, and taxation, among others. This governing body would then tap one of its members to serve as mayor to preside over the commission's deliberations. Court challenges compelled a crucial revision—all commissioners would be elected, not appointed—and once resolved, the Galveston commission set about reconstructing the fragmented port city.[12]

Many of Texas's major and minor cities quickly adopted the commission form of government as an efficient, effective, and impartial antidote to what its elite proponents believed was an old-time, ward-boss, patronage-driven politics that had dominated their communities. Ousting immigrants and racial and ethnic minorities from access to the levers of power was part of the Galveston Plan's appeal for its proponents. Envious white reformers in San Antonio watched as Houston, Dallas, Fort Worth, El Paso, and then a slew of communities across the country, adopted this new governmental structure. They were certain that if they, too, could rewrite the local city charter, a brighter future awaited. Instead, Bryan Callaghan lay in wait. In 1905, after being out of the public eye for six years, he stormed back into office, campaigning against the idea of commission government and in favor of reduced spending, lower taxes, and fewer improvements. These goals were the inverse of what had defined his previous mayoralties, and perhaps were a well-calculated slap at business leaders who had earlier pushed him out of city hall.

For the next six years, Callaghan's used his powerbase to fend off the demand for a referendum on commission governance. Finally, in 1911, pressed by the San Antonio Commission Government League, he agreed to bring the issue before the electorate. The campaign was bruising. Reformers made it clear whom they were running against: "There cannot be any personal or political machine in a government that is governed by a board of directors directly responsible to the people instead of the boss or bosses."[13] Callaghan's allies offered a sharp rejoinder, alleging that local do-gooders had imported faddish ideas and fancy outsiders "to San Antonio to tell her citizens how to run her government."[14] The anti-commission faction managed to win by a mere 168 votes (only a cynic—and there were plenty of them—would have suggested that Callaghan influenced the outcome by appointing police officers to serve as poll watchers).[15] Although Callaghan would win his ninth term in office the next year, he would die shortly thereafter from complications of acute kidney disease. With the death of this ardent champion of the poor and working class, a group for whom ward politics had offered a voice in public affairs (and a perhaps a job in city government), the anti-Callaghan faction pushed once again to enact the commission government. In 1914, they succeeded, but with a twist that Callaghan would have appreciated. Now-mayor Clinton Brown, a businessman and scion of the founders of the Alamo National Bank, "had no intention of abandoning patronage as long as he controlled city government. Nor did he intend to dilute his powers," argue scholars John Booth and David Johnson. Brown presented voters with a revised city charter that did not include civil-service hiring based on merit, did not create a ceremonial mayor but gave the position veto power over expenditures, and did not include term limits. Brown had co-opted the reform movement to such an extent that from James W. Tobin's election as mayor in 1919 until 1954, when another generation of crusaders, this time waving the banner of the Good Government League, instituted a council-manager form of government, San Antonio was run by series of Callaghan-like machines.[16]

Yet as people-oriented as Callaghan was, and as systems-focused as his rivals were, neither he nor they paid much attention to nature and its power to disrupt human affairs. A reminder of its punishing force came in the fall and early winter of 1913. During the first two days of October, an El Niño-energized storm pounded the San Antonio River watershed, dumping more than nine inches of rain that raced down bone-dry Olmos Creek, barreled into a fast-running San Antonio River, and then rose up and over its banks. Four people drowned, as the rank, debris-filled waters surged down streets, sucked houses into the torrent, weakened bridges, inundated cellars and ground-floor lobbies, and swirled through dry-goods stores, sending shoes, hats, and clothes downstream. Police and firefighters, along with a large detachment of soldiers from Fort Sam, pulled people out of the roiling floodwaters, provided food, shelter, aid, and comfort. Their quick work earned local officials praise, who credited them with preventing an even greater loss of life.[17]

Those same officials, notably Mayor Brown, dusted off an old plan to build what he dubbed a "Great Dam" across Olmos Creek at the point where it cut through the low hills just north of its confluence with the San Antonio River. The concept for this flood-control project had first surfaced in the scouring aftermath of the 1865 flood, and would pop up in subsequent years whenever San Antonio went underwater. Each time, after political leaders and business elite had proclaimed the concept's virtue, the land had dried out and they let the opportunity drift away. In 1913, however, nature sent a second wave of storms that should have driven home the message that if San Antonio wanted to continue to occupy a floodplain, it needed to take steps to protect itself. In late November and early December, having barely cleared the city's mud-slick streets, retrofitted its compromised bridges, and restocked business inventories, San Antonians endured another series of earth-saturating rains. During the early morning hours of December 4, a deluge triggered another rush of water downstream. Sweeping down the

Flooding on St. Mary's Street looking south from E. Houston Street, December 1913. *General Photograph Collection, UTSA Special Collections.*

same path as the October surge, this latest flood—whose volume was estimated at 8,000 cubic feet per second—raced across the central core, leaving wreckage strewn in its befouling wake. The next October, yet one more blockbuster storm crashed overhead, with five inches of rain slashing down over a three-hour period. Centered over the upper reaches of the narrow-banked and brush-choked west side creeks—the Zarzamora, Alazán, Apache, and Martínez—the rampaging waters slashed through the abutting and densely settled neighborhoods, undercutting houses, outhouses, and corrals, and turning the wooden debris into battering rams that hammered other structures, trapping residents and animals. Nine people drowned, including a day-old baby and its family, a horrific loss.[18]

The triple whammy of floods generated the now-standard public commitments to building a dam, reconstructing the San Antonio River's serpentine course through downtown, and a studied silence about the havoc waist-high floodwaters brought to the heavily Mexican American west side. Hoping for a quick fix, the city asked an engineer working on the local sewage system to offer an assessment of San Antonio's flood-control needs. The resulting report recapitulated the most pressing needs and suggested that the city hire civil engineers whose expertise lay in flood control to produce a much-more complete evaluation. However sensible that latter suggestion, and however modest the fee for this report's completion—$800—city hall got cold feet after a local newspaper challenged what it considered a gross expense to taxpayers.[19] Not until 1920 would elected officials brave potential censure and commission a hydrological analysis of the river that included a series of ever-more expensive solutions to San Antonio's downtown flooding problems. After six months of study, in December 1920 the Boston firm of Metcalf & Eddy laid out its recommendations. It argued for the construction of a dam to retain floodwaters from Olmos Creek; the straightening of the river at six key points to decrease water jumping the bends and turning streets into streams; and the deepening, widening, and armoring of the channel through the central core so that it could carry upwards of 12,000 cubic feet per second. The price tag of $4 million may have seemed steep to San Antonians used to doing things on the cheap, but Metcalf & Eddy, which had a good sense of the community's penny-pinching proclivities and its historical amnesia, warned against inaction. "We doubt the citizens realize the ruinous loss which would result today with the present condition of the river channels, from such a flood as of that of a century ago (1819). When such a flood will recur, no man can say. But that it will recur is certain." To do nothing was to court disaster, the engineers insisted; waiting would prove a death knell. "We counsel the wisdom of pushing this work . . . while the memory of recent floods is vivid, lest the public mind relapse into inaction in a false sense of security when the inevitable flood shall come. We urge that your citizens shall remember that this flood is just as likely to come next year as at any other time." Nine months later, in September 1921, a century-level flood ravaged the city.[20]

At the same time that community leaders engaged in the most serious discussions of how to control the river, others sensed that these in-depth conversations opened the way for an aesthetic makeover of its banks, a beautification initiative that would benefit the local tourist economy. These river reformers imagined a series of tree-shaded paths that would wander along the river, green esplanades of flowering shrubs and plants." New shops at street and river levels would cater to those strolling down sidewalk and pathway. With boats plying the languid waters, and colorful lights reflecting on their rippled wake, this pedestrian environment would lure in local and visitor alike. These dreams gained an important political ally after the installation of the commission form of government: one of its member's oversaw the river and parks, giving these resources a strong voice in the city's deliberations and expenditures. This elevation proved useful when citizen-activists urged Commissioner George Stuckey to support their conceptions of the riverway's prospects. His support, which included providing labor and funds for the construction of walks and other amenities, while not solely driven by his instinct to curry voters' favor, contained its share of back-scratching reciprocity. Yet while these beautifying efforts gained national attention, there were limitations to what could be accomplished in the absence of an upstream dam. Floods in 1919 and 1920, while cresting below street level, did considerable damage to riverside landscaping and lighting, a swift reminder of the ephemeral nature of these well-intentioned efforts. Another, more politicized challenge came in March 1921 as the city's commissioners, in response to Metcalf & Eddy's final report, committed money to rip out any obstacles—trees, bushes, shrubs—that would impede the movement of floodwaters through town. When the public got wind of the impending changes, those most committed to maintaining the river's park-like character immediately called a public meeting to lambaste Mayor Sam Bell and Parks Commissioner Ray Lambert. Scurrying for cover, the city commissioners scuttled their budgetary commitment to clear-cutting the river's banks. This tension between the aesthetic and utilitarian approaches to the river's management would not dissolve until the construction of the Olmos Dam in the mid-1920s.[21]

Floodwaters compromised another crucial element of the built landscape: the movement of people, goods, and services. Although animal-drawn wagons and carriages could roll through high water, those means of conveyance were diminishing in number and importance by the late nineteenth century. Replacing them were trolleys and streetcars, the first version of which appeared in 1878, a project that marked a dramatic shift in the way the oft-flood-wracked city would develop. The inaugural line, which private, outside invest-

1916 San Antonio River beautification. *Courtesy of the Edwards Aquifer Website by Gregg Eckhardt (http://www.edwardsaquifer.net/).*

ment financed, rolled from downtown uphill to San Pedro Springs. Its route seemed odd at the time, a reporter mused ten years later, for the "streets through which the line traveled were in miserable condition; San Pedro Avenue, now, one of our finest streets, was covered with brush and small trees, and the cars virtually ran over a prairie, there not being a single house on the street." That emptiness changed quickly, as the owner, August Belknap knew it would, given how streetcars had opened up new residential space in his native New York City. The investment in the trolley's lines was a loss leader: the sale of adjacent land paid huge dividends. "Large lots, which could be bought for $5 and $10, immediately jumped in price to from $100 to $500. A rush was made by citizens to that locality, and very soon dozens of residents were constructed on the avenue." As Belknap extended his routes along a once-deserted Houston Street, then linked cross-city railroad depots—Sunset station on the east side with the west side's International & Great Northern, and subsequently the Aransas Pass terminus on the south side—he created a grid that facilitated development between and around these economic nodes. With every new mile, commercial structures or housing sprouted up, benefits that came with deficits. A "big howl went up from the livery stable men, as [the streetcar] was making big inroads into their business," but well-off property owners, like "residents all along the line feel under great obligations for the comforts which they now enjoy, by being able to reach their homes in all kinds of weather without any inconvenience."[22]

The pattern was set: this new conveyance, the streetcar, which only those with sufficient disposable income could afford, intensified the shakeout of the city's residents according to class, race, and ethnicity. In this new San Antonio, which by 1910 was home to more than 96,000 people, some rode, while others walked. The streetcar-driven escalation in the price of land and housing reinforced that distinction in mobility, as did the expansion of these lines ever farther from the downtown core. By 1917, nineteen streetcar lines crisscrossed and advanced the city's now-sprawled extent. As they extended the city's developed margins, these lines' outlying terminuses told another, related story about spatial segregation. Each was located on high ground. One halted at Prospect Hill on the west side. On the north side, streetcars made possible suburban subdivisions named Beacon Hill, Tobin Hill, Monte Vista,

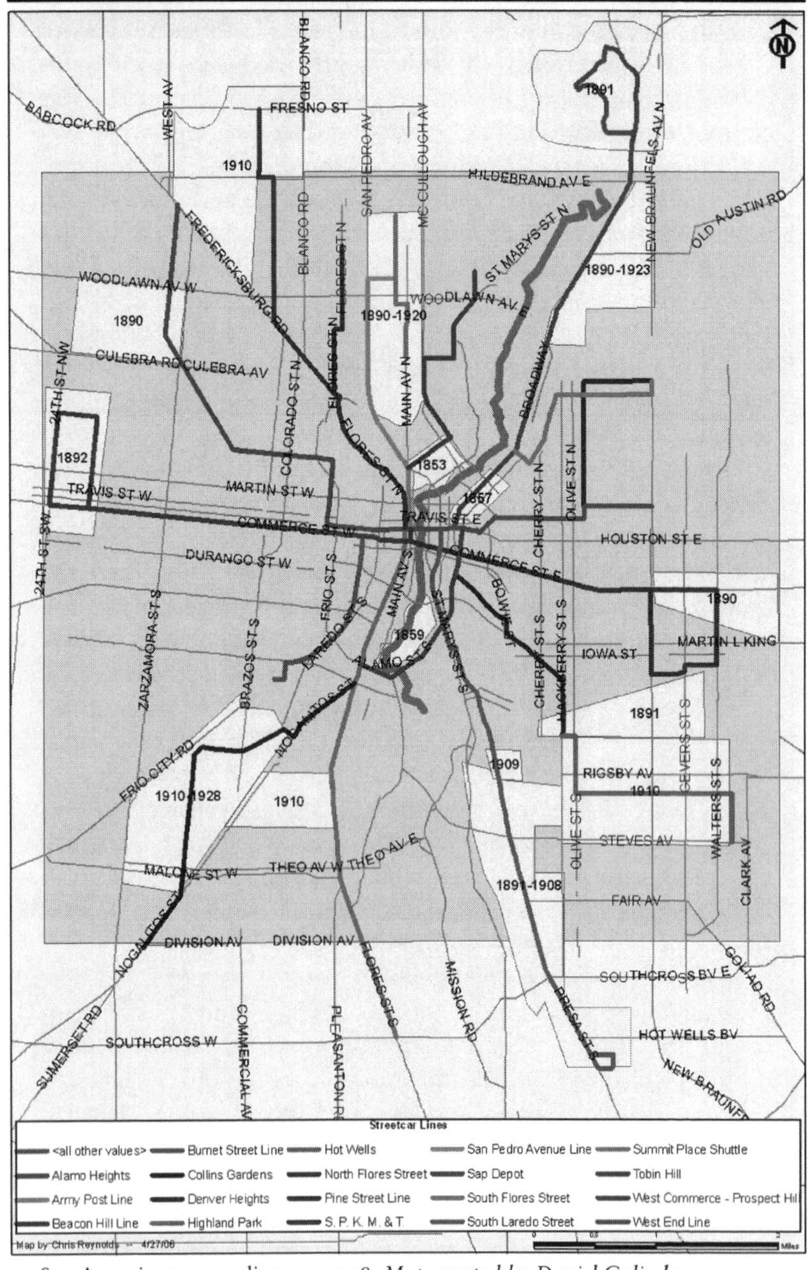

San Antonio streetcar lines, c. 1918. *Map created by Daniel Galindo.*

and Alamo Heights. On the east side, tracks ran up to Government Hill (home to Fort Sam Houston), and another other rose up to Denver Heights. The human geography converged with the physical. Those who could pay the increased prices associated with this elevated acreage, did so. Those who could not, crowded into the as-yet unprotected floodplain. Bluntly put, the Anglo elite stayed dry; poor Tejanos, blacks, and whites did not. The city would not begin to address this intertwining of social injustice and environmental disadvantage until west side activists forced it to do so in the last third of the twentieth century.[23]

7
All Quiet on the Southwestern Front

On Armistice Day 1918, a huge crowd gathered in downtown San Antonio, blocking Alamo Plaza and Houston Street, to witness a spectacular airshow: a "circus" courtesy of thirteen pilots from Kelly Field. The spectators' mouths gaped as the planes soared overhead, held "breathless with the most daring stunt flying ever exhibited over this city." On their initial pass over the Alamo, the thirteen planes flew in close formation and dropped flowers in honor of those who had died during World War I. That pacific gesture gave way to a set of stunts designed to imitate the war itself. After reaching an altitude of five thousand feet, the pilots put their planes into steep dives "until the wires and motor screamed and roared beyond description." At what seemed like the last moment, each pilot hauled back on the stick, pulling the plane straight up in the air and once it regained enough altitude, turned the plane "completely over, making a loop in the air." Then "SE 5's, like hawks, darted across the sky pursuing each other in dives, turns, wing-overs, reversements, and other battle tactics familiar to those in action overseas, but revelations to people who never before have seen real pursuit flying." Just as revelatory was a sham dogfight between two airplanes high above the plaza. Those on the ground watched in thrilled horror as the pilots demonstrated through a series of "dives and turns of every twisting and confusing nature known to modern trained airmen" how to get the best of their foe. It was, a reporter asserted, "an air battle exactly as it would take place between two hostile air craft on any battlefront," minus machine guns "spurting deadly bullets at each squeeze of the trigger." The thunderous airshow, in the end, delivered a "new education in flying to the residents of this aviation city."[1]

No one could claim that San Antonio was an aviation hub a mere eight years earlier when the U.S. Army's first pilot, Lieutenant Benjamin D. Foulois, arrived in the city by train. No acclaim greeted his arrival, either, and with good reason. He was not really a pilot, the army had no idea how to integrate airplanes into its ground-fighting strategies, and it had reluctantly encumbered $25,000 to purchase its first aircraft from the Wright Brothers at the insistence of then-President Theodore Roosevelt, who believed it had potential as a deadly tool of war.

Why San Antonio? In retrospect, it makes sense for reasons of climate and space. South Texas enjoys upwards of three hundred sun-filled days and the region's flat coastal plain is ideal for pilot training. At the time, however, the army could not figure out where else to conduct its aviation experiments, so it dispatched Foulois and a maintenance crew to Fort Sam Houston. When he stepped out of the railcar at Sunset Station in February 1910, among his baggage was several large crates carrying the disassembled Aeroplane No. 1. Foulois, whose orders were succinct—"take plenty of spare parts and teach yourself to fly"—worked with his detachment of enlisted men to build the American armed forces' first hangar and to transform a cavalry drill field to double as his runway. On March 2, 1910, he managed to take off in the long-winged bi-plane, fly around the parade ground, and then force the plane down for an emergency landing when a fuel pipe cracked. That flight was historic on two counts. It was Foulois's "first solo, landing, take off, and crash," as he joked to his superiors, and it was the first military sortie over San Antonio.[2]

However accidental it may have been that San Antonio became ground zero for the army's tentative experiment with aerial warfare, there was no mistaking what happened seven years after that Foulois reported for duty. The Alamo City became the epicenter of military aviation in the United States, as became evident in the important contributions it made to the nation's preparedness for World War I. Although thousands of miles away from the ghastly trenches in eastern France, buffered by the Atlantic Ocean and much of the North American continent, and therefore largely immune to the vicissitudes of a devastating conflict that claimed upwards of twenty million lives, the Alamo City and its hinterland helped place the nation on an aggressive wartime footing.

The initial American response to the brutal war that erupted in August 1914 between the Triple Entente (Britain, France, and Russia) and the Central Powers (Germany, the Austro-Hungarian Empire, and Italy) was to remain neutral, a strategy that President Woodrow Wilson affirmed over the next two-and-a-half years. Although Wilson's position dovetailed with public opinion, neutrality became increasingly difficult to maintain as the German and British navies used their competitive advantages—submarines and ships, respectively—to destroy the other's capacity to fight. In the middle of this intensifying battle was the American merchant fleet. As it plied the North Atlantic carrying food, fuel, and weapons to the belligerents, the vast bulk of which ended up in French or British ports, the Germans began to target this vital supply line, and their torpedoes began to take a toll on the flow of cargo.

It was the German attacks on passenger liners, however, notably the 1915 sinking of the *Lusitania*, which killed 1,198, including 128 from the United States, which by early 1917 intensified American sympathies for the Triple Entente's cause. So did the Zimmermann Telegram, a January 1917 coded telegram sent from Berlin to the German ambassador to Mexico. In it, Foreign Minister Arthur Zimmermann promised that if Mexico sided with the Central Powers, a victorious Germany would seek restoration of one-time Mexican territory in Texas, New Mexico, and Arizona. The subsequent national uproar was particularly loud in the Lone Star State. The *San Antonio Light*, one of the city's major newspapers, vowed that should a German-Mexican army push north it would be met by Texans willing to fight to the death.[3] In response to the intensifying pressure, on April 2 President Wilson called Congress into "extraordinary session because there are serious, very serious, choices of policy to be made, and made immediately, which it was neither right nor constitutionally permissible that I should assume the responsibility of making."[4]

Wilson's brief for why the United States now should take up arms was carefully reasoned, as befit a former university teacher and president. Because "the German use of submarines violated international law," the president asserted, and because their destructive force was aimed at the "wanton and wholesale destruction of the lives of non-combatants . . . engaged in pursuits which have always, even in the darkest periods of modern history, been deemed innocent and legitimate," it had become manifestly clear that neutrality "is no longer feasible or desirable."[5] Under these changed conditions, armed confrontation was the only answer: "the world must be made safe for democracy."[6]

Wilson's words found favor in San Antonio. The *San Antonio Light* cheered: "The war between the United States and Germany which all farsighted people have long beheld as a possibility, and which all right-minded people have conscientiously endeavored to avoid, has come." This conflict erupted, the newspaper assured its readers, because "our good intentions have been misconstrued, our patience has been abused, our forbearance has been outraged." Now it was up to Americans to "prove to the world that, although a peace-loving people, we are still the children of those who . . . have never failed . . . to finish in their own way those things to which they have set their hand."

The newspaper's audience no doubt shared these patriotic sentiments, because San Antonio long had been at the center of armed conflict, from its distant past as a fortified outpost on the northern frontier of New Spain to its more recent experience with Fort Sam Houston training troops to defend the U.S. empire in the Pacific and Caribbean. These latter, and more recent overseas engagements proved formative in a larger sense: they were rigorous testing grounds for the American armed forces and their equipment in the run up to World War I. The most immediate test came as a result of the 1916 Punitive Expedition, under the command of General John "Blackjack" Pershing, then headquartered at Fort Sam Houston. President Wilson had ordered this strike force to the U.S.–Mexico border to prevent Mexican guerillas from attacking across the Rio Grande. The Punitive Expedition did not stop at the border. With winged support from the First Aero Squadron, also stationed in San Antonio, it pushed deep into northern Mexico in a futile attempt to capture revolutionary Francisco "Pancho" Villa, whose troops had raided Columbus, New Mexico, in early March 1916. As part of this action, more than 150,000 members of the National Guard were called up, a large portion of whom were shipped to San Antonio to receive basic training at Camp Wilson and practice gunnery and tactical maneuvers at Leon Springs Military Reservation (later Camp Bullis). Although the expedition failed to take Villa into custody and its troops were mustered out of federal service in late 1916,

Kelly Field, San Antonio in the 1920s. *John J. Pershing Collections, Library of Congress Prints and Photographs Division, Washington, D.C.*

these newly hardened soldiers and officers had learned invaluable lessons that they would put to good use when they were remobilized only months later. On April 6, 1917, the U.S. Congress issued a formal declaration of war with Germany: "Whereas the Imperial German Government has committed repeated acts of war against the Government and the people of the United States of America; Therefore be it Resolved . . . that the state of war between the United States and the Imperial German Government which has thus been thrust upon the United States is hereby formally declared."[7]

The long-awaited declaration was an economic boon to San Antonio. Its extensive history as a military service center and staging ground, a mission that the U.S. Army had expanded with the 1915 construction of its first aviation facility, Dodd Field, on the grounds of Fort Sam Houston, insured a remarkable influx of trainees for the duration of the war and an steady infusion of federal dollars into the city's coffers.

Bases old and new dotted the local map. The Leon Springs training ground, which before the war covered 17,000 acres, grew to more than 32,000 during the war, with the army spending $1,350,000 on the construction of its facilities and services.[8] Camp Wilson was renamed Camp Travis, and overnight became a self-functioning city replete with 1,400 temporary buildings built in three months and in which more than 100,000 soldiers—approximately 10 percent of all Americans who served in Europe during the war—were housed, fed, and taught. Because Dodd Field could not accommodate the army's escalating demands for flight-ready pilots

Camp Wilson, Fort Sam Houston, San Antonio. *General Photograph Collection, UTSA Special Collections.*

and ground crew, the Department of War leased 700 acres south of the city for Kelly Field and within a year added another 1,800 acres. This field's rapid expansion was necessary, given its mid-boggling training schedule: on Christmas Day 1917 more than 39,000 men were stationed at the airfield, and the next month, despite shipping out 15,000 cadets, another 47,774 recruits had arrived. Cumulatively, notes historian Thomas A. Manning, "Kelly soldiers organized approximately 250,000 men into aero squadrons during the hectic months of 1917 and 1918. The Enlisted Mechanics Training

Department turned out an average of 2,000 mechanics and chauffeurs a month. Most of the American-trained World War I aviators learned to fly at this field, with 1,459 pilots and 398 flying instructors graduating from Kelly schools during the course of the war."[9] Among these was a local boy made good: Edgar Tobin, a member of Captain Eddie Rickenbacker's "Hat in the Ring" air squadron and a highly decorated ace.

Kelly Field's explosive growth was such that yet another airbase, Gosport (later Brooks) Field, had to be carved out of the city's south side. On its 1,300 acres, pilots were trained to fly balloons and airships. Civilians pitched in as well: Marjorie Stinson, who with her daredevil-pilot siblings, Katherine and Edgar, had trained fliers at Fort Sam Houston before the war, continued doing so once hostilities began, operating out of Stinson Field, the city's municipal airport. Soldiers marched, airplanes whizzed overhead, rifle fire crackled, and howitzers thundered: however far removed San Antonio was from the war's major battlefields at Verdun, Ypres, and the Somme, the city must have sounded and smelled as if it was in the very midst of war.

Walking its downtown streets would have confirmed as much. Uniforms were everywhere; military vehicles crisscrossed the city between its many bases, depots, and camps; and entrepreneurs eager to cash in on wartime spending were in abundance. Some visionaries hoped that the war would expand the city's small industrial base. What better item could it manufacture than airplanes? These ambitions made sense but inept leadership, divisive politics, and the failure nationally for postwar aircraft production—military and civilian—to take off, scuttled the idea. Yet out of this civic activism emerged a greater emphasis on military spending in San Antonio's economy and thus a greater dependence on its appropriations. In times of conflict, the citizenry would discover that could be a very good thing; but in times of peace, as federal funding dwindled, so would the community's fortunes.

Despite the unreliability of federal funding, the city's population grew: in 1910, 96,000 lived in San Antonio and at the close of the war it was home to more than 160,000. These figures do not factor in the waves of temporary residents who flooded into town for basic training or advanced instruction. Few of these newcomers had ever weathered anything like the region's withering heat

Marjorie Stinson. *Records of the War Department General and Special Staffs, 1860–1952, American Unofficial Collection of World War I Photographs, 1917–1918.* Source: https://catalog.archives.gov/id/533715 [Accessed Mar. 1, 2018].

or sticky humidity, its furious thunderstorms, or the bone-chilling "blue northers" that screamed down the Texas plains. One pilot remembered "arriving at Kelly Field after withstanding a long, hard trip, when food had been given out thirty-six hours earlier, with great anticipation of becoming a great aviator and of making fame by bombing old Hun "Bill's [Kaiser Wilhelm] palace," only to snap to attention when he found himself "lined up in front of a row of tents . . . feet in black mud and the wind blowing a gale, trying to obey the orders—'prepare for inspection.'"[10] Others chaffed at the constant dust, rain, and sandstorms that swept across the open ground, making it "impossible to keep clean"[11]

Just as bewildering to these newcomers were their close encounters with San Antonio's notorious red light district, a vast warren of bars, gambling emporia, and brothels located just west of the central business district. It reminded an aviation cadet of what he imagined a "roaring, wild west town" would look like. The city "seemed to abound in saloons that were decorated with extremely long cow horns on the walls, and with glass enclosed, realistically preserved habitants of the state, including coiled rattlesnakes and hairy tarantulas, to greet the visitor's startled eye from many nooks and crannies."[12] For recruits, San Antonio was an unusual place in a disorienting time.

Some locals felt similarly baffled. Those of German ancestry, many of whose families had arrived in the region in the 1840s and contributed significantly to the community's cultural life, economic energy, and political affairs, became targets of xenophobic animosity. Anti-German boycotts hit their stores, banks, and breweries. King William district, "home of the most successful families of German heritage, was called 'Sauerkraut Bend,'" and the city also renamed the neighborhood's eponymous thoroughfare to "Pershing Street, after U.S. General of the Army George Pershing." A different form of prejudice awaited Morris Kallison, son of a prominent Jewish mercantile and ranching family in San Antonio. During his artillery training at Fort Sam Houston, he became skilled in handling howitzers and mortars, but he discovered that he also would have to fight against anti-Semitism. "Assimilated as a teenager into San Antonio's diverse community," Kallison, a trained pugilist, did not back down when a fellow soldier unleashed "a vicious anti-Semitic epithet" his way. "With the strength and skills honed while boxing as 'Kid Morris,' Kallison pummeled the bigot."[13] The war unleashed an ugly nativism on the bases and in the streets.

Then suddenly hostilities ceased. The warring parties signed an armistice on November 11, 1918, a mere seventeen months after the formal American declaration of war. The news touched off enthusiastic demonstrations from coast to coast; and ticker tape swirled. San Antonio's rival papers dueled with one another over the most punched-up headline. "EMPEROR ABDICATES," the *San Antonio Light* trumpeted; blared the *San Antonio Express*: "GERMANY GIVES UP," with subheads claiming "Everything for Which America Fought is Won." Yet back at Kelly Field, at least initially, the trainees were in no mood to celebrate Germany's "Absolute Surrender."[14] When word reached them, there "was dead silence for about 30 seconds," one trainee recalled, and then "the whole barracks sat up and started cursing at the floor, damning the Germans for having lain down on the job before they had a chance to get over there and prove their merit as aviators."[15]

Their commander in chief knew better than to prolong the bloodletting. More than 100,000 Americans had lost their lives, another 200,000 had been injured, and the nation's wounds—social, political, and economic—went deeper still. That is why President Wilson's "Thanksgiving Address," delivered within days of the armistice, only spoke of pride and of hope: "Complete victory has brought us, not peace alone, but the confident promise of a new day as well, in which justice shall replace force and jealous intrigue among the nations." Lauding our "gallant armies" and the "righteous cause" for which they battled, Wilson knew that the American Expeditionary Force had "nobly served their nation in serving mankind." Their accomplishments were a source of great joy: "We have cause for such rejoicing as revives and strengthens in us all the best traditions of our national history. A new day shines about us, in which our hearts take new courage and look forward with new hope to new and greater duties."[16]

Many local Mexican Americans shared Wilson's dream. Thousands of Mexican Americans in San Antonio and across the Southwest had enlisted to prove their loyalty, and, having fought honorably, believed they could demand political change "to make democracy a reality and reality democratic."[17] Veterans returning to San Antonio and elsewhere quickly realized that many of the pre-war discriminatory practices continued. They discovered "that they were not served drinks," for example, "and were told 'no Mexicans allowed.'" This bigotry raised a galvanizing question, one soldier recalled. "What are we, Mexicans or Americans? The world war taught us a lesson. We had thought we were Mexicans. The war opened our eyes . . . We have American ways and think like Americans." To lay claim to political power, achieve greater economic opportunities, and secure social equality, they realized that they would need to organize. The Order of the Sons of America, founded in San Antonio in 1921, grew out of this realization. As John Solís, one of its founders, argued: "we didn't have anything in 1921 [so] we decided to organize our people and try and better the condition of Mexican Americans here in San Antonio and Bexar County." It would take a half century of organizing before local activists could secure significant gains at home, but the Great War helped stimulate this drive for social justice.[18]

It also set the stage for African Americans to challenge local racial inequities. The National Association for the Advancement of Colored People (NAACP), founded in 1909, gained its first Texas chapters during the European conflict. That war had its horrific domestic counterpart—a deadly spike in lynch-mob violence. In reaction, NAACP organizers fanned across the state, selling Liberty Bonds and promoting equal rights under the law. Two years after El Paso organized a chapter in 1915, San Antonio followed suit, and another twelve across the state opened the next year. As the driving force behind the San Antonio chapter, John A. Grumbles was instrumental in linking the need to advance the cause of African Americans to the war effort. He invited General Pershing to speak at NAACP meetings, and rallied blacks and whites against *The Loyal Spirit*, a patriotic magazine that denounced the presence of "Negro soldiers" in the city's military training facilities. Grumbles built up such a drumbeat of opposition that the periodical disappeared from San Antonio. Grumbles was just as central in securing a substantial number of civilian jobs on local bases, including upwards of three hundred seamstress positions for black women. It is not surprising, then, that by 1919 Grumbles had grown the local NAACP chapter into the state's largest. Many of these impressive gains in employment evaporated at war's end, and with Grumbles's sudden death in 1922 the African American community lost a key powerbroker.[19]

The nation itself lost an opportunity to change the world. President Wilson's generous Fourteen Points for postwar reconstruction on the international front, which included the creation of a League of Nations to guarantee political sovereignty and territorial independence, ran into considerable opposition. His European allies, especially the imperial powers of Great Britain and France, worked assiduously to undercut his ambitious goals, as did the president's congressional opponents. The domestic front was filled with many other troubling developments. The United States was rocked by a freefalling economy, the deadly Spanish Flu pandemic, "Red Scare" deportations and imprisonments, the rise of the Ku Klux Klan and a nasty xenophobia, a brutal crackdown on labor unions, and vicious race riots that ripped through more than twenty-five cities, including Chicago, East St. Louis, and Tulsa. The war's end brought little peace.

San Antonio experienced its share of postwar social turbulence, an unhappy state of affairs that clouded the first years of its bicentenary. As its military infrastructure shut down, commercial activity went into a tailspin, leaving the city's newfound workforce with fewer employment options. That dispiriting situation stabilized by the mid-1920s, only to worsen with the onslaught of the Great Depression, calling into question the wartime sacrifices so many of its residents had made—the Liberty Gardens they had planted; the natural resources they had conserved; the War Saving Stamps they had purchased; and, most devastatingly of all, the loved ones they had lost.

These deep-seated and understandable concerns aside, the city of San Antonio had directly benefited from the First World War. It absorbed millions of dollars in federal military spending to train hundreds of thousands of troops, building up the community's capacities while helping, as one early wartime poster put it, to "Beat Back the Hun."

8
Turbulent Twenties

Across the nation, recovery after the Great War was slow and uneven, marked by inequity. This was certainly true in San Antonio, where a sharp decline in military spending after the Armistice led to the closing of many local bases and facilities. Those merchants who supplied necessary goods and services to the U.S. Army's armory, its training camps and airfields were hit hard, as were the thousands of civilians who were let go after laboring in close support of the war effort. The city's economy shuddered, capital was scarce, and unemployment spiked. What helped the community weather the storm were the agriculture and ranching sectors that dominated its hinterland and for which San Antonio served as the provisioning hub. Of aid, too, was the return of tourists, who arrived by train, and, increasingly, automobile, to visit the city's eighteenth-century relics—the Alamo, Main and Military Plazas and the four crumbling missions scattered down river. Every dollar helped.

That other people's money maintained the local economy was also problematic, a symptom of a larger failure that for decades had dogged the community's economic fortunes. Unlike its regional rivals, Dallas and Houston, San Antonio had not yet developed a core group of entrepreneurs and boosters willing to risk their time, energy, and fiscal resources to build a more robust urban marketplace. Compared to its competitors to the north and east, which had invested heavily in transforming their cities into banking centers, industrial nodes, and transportation hubs, San Antonio seemed reluctant to pursue such opportunities. During the buildup to World War I, for example, two local entrepreneurs proposed that the city's commercial sector expand its engagement by marshaling the necessary private and public capital and building a deep-pocketed partnership with the Army

Air Corps to construct military aircraft. Taking advantage of this new technology would have been a natural extension of the community's longstanding commitments to the War Department and would have distinguished it from its urban peers. The newly formed Chamber of Commerce passed on the project. In the early 1920s, it would also fail to take advantage of another opportunity. Agricultural interests in South Texas, frustrated by the lengthy travel required to ship perishable goods to Houston, urged San Antonio leaders to join them in underwriting the construction of a railroad between the valley and the city. Their appeal gained little traction. The deal fell apart because nothing in the "collective wisdom of San Antonio's entrepreneurs had taught them the importance of cooperative action for a common good," write two historians of the city. Because in the past urban growth "had occurred without assuming risks or employing collective effort," community leaders figured it would in the future. One relevant consequence of this assumption was that as Dallas and Houston expanded rapidly in the postwar years, so did their populations: San Antonio, which in 1920 was the largest city in Texas, by 1930 had slipped to third.[1]

It took an unprecedented disaster to mobilize the business elite, a disaster so profound that it partly explains San Antonio's laggard growth over the 1920s. That said, even as the city's powerbrokers acted quickly in the wake of the devastating flood of 1921, their actions were narrowly framed, self-serving, and punitive in some their long-term social consequences. Late in the evening of September 9, San Antonio went under water. So did the other communities lying along the Texas Spring Line, including New Braunfels, San Marcos, and Austin. The staggering volume of water that came crashing down resulted from a slow-moving hurricane that several days earlier had churned ashore in northern Mexico. As it spun over the Rio Grande, it dumped upwards of six inches of rainfall on Laredo, sinking low-lying neighborhoods. Pressing north, it cycled along the Balcones Escarpment, where the storm unleashed its full fury. Thrall, a small town in Williamson County, recorded a staggering 38.21 inches of rain in a twenty-four-hour period and Austin got 18.23.

These fixed-site numbers do not tell the whole story. Upstream from each of these communities, heavy rains were also slashing down into the folds of the Edwards Plateau, rocketing off the pan-hard ground into the gullies, streams, and creeks that fed into the Little Colorado, Comal, Guadalupe, and San Antonio Rivers. These roiling surges quickly blew over their banks. Farms, ranches, and feedlots were inundated. Houses and barns were swept off their foundations and careened downstream. Uprooted trees became battering rams, slamming into buildings, tearing up transportation infrastructure, and hammering bridges.

Property damage was intense, but the loss of life was more so. In Taylor, eighty-seven people died, and another six in its home county of Williamson. Six were swept away in Travis County, and the statewide total was 215. By all measures, the 1921 floods were the most devastating in the history of the Lone Star State. It was also the most destructive San Antonio had ever endured, a significant claim given its flood-prone nature. The eighteenth-century Spanish planners had platted the early community inside a flood basin, locating its streets, plazas, and residential areas between two river systems, the San Antonio to its east and San Pedro Creek to the west. As a result, San Antonio frequently foundered. Until its roads were hardened in the late nineteenth-century, even light rains had turned streets into a mud-sucking mess. In some neighborhoods, homeowners had jacked up their houses above street-level to mitigate routine inundations. Yet no one in the city was willing to spend the money necessary to protect the central core, and the cost of this indifference and dithering became clear during the blockbuster floods of 1819 and 1865, when local waterways raged through the town, trapping residents and devastating homes, churches, and commercial buildings.

As destructive as those nineteenth-century floods were, the 1921 event was far worse. Although only seven inches of precipitation fell to the north of the city, much of it came during a concentrated downburst in the dead of night. Upstream ditches and creeks rose swiftly and then slammed into San Pedro Creek and the San Antonio River, which could not contain the turbulent waters, sending wave after wave down alleys and avenues. The peculiar nature of the city's siting proved disastrous.

The *Austin American* caught some of the horror that then occurred in San Antonio. On the west side, home to a large Latino barrio, and through which Alazán Creek threaded, the usually "placid rivulet of water became a rushing torrent in less than half

1921 USGS flood map. *Courtesy of the Edwards Aquifer Website by Gregg Eckhardt (http://www.edwardsaquifer.net/).*

an hour," rising more than "eight feet in approximately twenty minutes." Almost immediately, the small homes and shacks near the creek bed began to float off their foundations," the newspaper reported, "and it was a barrage of these that hurled themselves against the International & Great Northern trestle. Their combined pressure "soon cracked that structure in the middle, which pushed itself against a second trestle that broke shortly after under the strain." Most of those who died—and the final tally was more than fifty—did so in this compact neighborhood, San Antonio's poorest.

Finding their bodies was not easy, and for the rest of the week able-bodied civilians and soldiers from Fort Sam Houston took up posts on the remaining bridges and riverbanks to spot, and then retrieve, corpses entangled in the sodden detritus.

Downtown suffered little loss of life, yet the damage was considerable due to the density of the built environment. Olmos Creek, which cut through the city's north and east, had powered over its banks around midnight, rushing into adjacent neighborhoods and urban parks. When its crest smashed into the San Antonio River, it forced a five- to ten-foot wall of water through the central business district. Most of the bridges that crisscrossed the San Antonio River gave way, their foundations undercut by the swirling water's erosive force or snapping before the pile-driving energy of trees, vehicles, and other flotsam. Commercial and civic buildings were flooded out; St. Mary's Street Catholic Church and others facilities with limestone footings proved structurally unsound and later were torn down. The Bexar County Courthouse was a mass of sodden paper, streetcar rail lines twisted like pretzels, and massive maze-like mounds of busted furniture, beams, street pavers, and dead animals choked the streetscape. The city lay in ruins, looking like a war zone.

It would take many months and four million dollars to clean up the mess. Even more expensive was the response of San Antonio's powerbrokers to the critical question of how to control future floods. They rallied around a plan calling for the construction of a retention dam across the Olmos Creek Valley to prevent floodwaters from coursing through the city. Their conception of the community was limited to its downtown, and when completed in 1927, the Olmos Dam's 1,900-foot-long structure did what it was supposed to do: keeping the core dry. The dam subsequently helped articulate land-use patterns within and redefine the spatial design of San Antonio. Its mere presence intensified the central core's already established economic functions by encouraging once-wary financiers to invest heavily in the city's post-flood skyline. A construction boom followed, producing some of San Antonio's finest and largest buildings, from Scottish Rite Temple (1924), Medical Arts Building (1926), the Smith-Young Tower (1929), and the Alamo National Bank (1929). The dam also made it possible to conceive and develop the River Walk. In 1929, landscape architect Robert

Flooding in San Antonio, 1921. *Courtesy of the Edwards Aquifer Website by Gregg Eckhardt (http://www.edwardsaquifer.net/).*

H. H. Hugman presented his initial plans for the site, dubbed the Shops of Aragon and Romula, to the city's political leaders. Without Olmos Dam and related flood-control infrastructure downstream, there would be few tourists strolling along the river's placid waters, tourists who would become one of the mainstays of San Antonio's modern economy. Taken together, these outcomes suggest that the Olmos Dam is the single most important public works project in San Antonio's modern history.

The dam was also flawed, in that the decision to build it depended upon a skewed distribution of public benefits in one of the poorest big cities in the United States. That is not what San Antonio's Anglo leadership announced when it floated a $3 million bond issue to construct the dam, insuring the central business district's bright future. It also voiced a commitment to establish a flood-control plan for the fatally waterlogged west-side barrios, lying far outside the proposed dam's protective zone. Outside observers urged the city to do more than simply build storm channels and sewers there. By sweeping away the concentrated tracts of "rude shacks, built in a hit or miss manner," the *Survey* reported, the 1921 flood offered San Antonio the best "opportunity for bettering the lives and sanitary conditions of the Mexican population."[2] The urban elite acted otherwise, focusing only on reconstructing the commercial core and ignoring the drainage, housing, and sanitation problems confronting those who lived on the west side. In August 1924, for example, at the same time the city council released millions of dollars to fund the Olmos Dam, it encumbered a mere $6,000 to cut brush along San Pedro and Alazán Creeks. This stark discrepancy in financial investment and flood-prevention technology, which would continue for the next fifty years, revealed the degree to which the management of San Antonio's flood waters, as with other thorny social issues, was channeled along ethnic divisions and class lines.

The dam's construction reinforced those same divisions in one of San Antonio's first automobile suburbs, Olmos Park, located just outside the city's northern limits. This elite enclave, with its substantial dwellings, tree-lined streets, and high-income residents, was deliberately segregated from the wider community and thus illustrates the political tensions and divisiveness that dominated San Antonio in the twentieth century. Its developer, H. C. Thorman, capitalized on the city's preexisting radial pattern of growth that streetcars and railroads had established since the late nineteenth century and that roads for automobiles quickly filled in and extended. Olmos Park's physical location enabled Thorman to exploit the presence of Olmos Dam. Prior to the dam's construction, he declared, "no facilities existed for some considerable distance for crossing between the east and west divisions on the north of greater San Antonio." The dam, with a road across its top, served as a bridge above the oft-dangerous Olmos Creek, and provided "a new crosstown thoroughfare linking Alamo Heights on the east with Laurel Heights on the west." Because Olmos Park lay between these two locations, it would benefit from new traffic patterns along the city's north side.[3]

The automobile was also responsible for the new subdivision's spatial form. Rather than reproducing the gridiron pattern characterizing most of San Antonio, Thorman designed Olmos Park to suit the car, and provide a sharp break—visual and physical—from the relentless right-angled streetscape. To enhance his property's "great natural beauty," Thorman lay down a series of wide avenues and drives that would wind among native oak trees and roll over the hills. An array of barriers then insulated the new community from its larger surroundings. To its east and north lay the Olmos Creek flood plain, land

Olmos Dam. *Library of Congress Prints and Photographs Division, Washington, D.C.*

the dam made uninhabitable; in the future, the flood plain would be turned into a city park. To the south lay a quarry, and on its western edge ran the Missouri Pacific rail lines. Each of these helped prevent the encroachment of undesirable development and, when linked to the centripetal force that the "parkway system" exerted on movement in the community's interior, set Olmos Park apart. Intensifying the new community's social exclusivity was a restrictive covenant that perpetuated racial segregation and maintained property values. It stipulated that Olmos Park was for whites only, as no property could be "sold, conveyed or leased to any person who is not of the Caucasian race." Violators would forfeit their title to their "particular subdivision of property," restrictions that Thorman vowed would "forever stand good."[4]

Olmos Park's covenant was arguably the most rigorous in San Antonio, making it difficult for any but a select few to live there. Among its first residents were oil magnates and cattle ranchers, as well as a growing group of San Antonio's commercial and business elite who hitherto had lived in areas threatened by the expansion of the downtown commercial sector, in the floodplain, or in neighborhoods that lacked restrictive covenants. An added benefit for those who relocated to Olmos Park was that they could avoid San Antonio's taxes and direct participation in its political affairs. Until the late 1930s, it was possible to maintain this isolationist posture because it dovetailed with the then-dominant city machine's interests.

Like other contemporary urban political machines, San Antonio's politicians bargained over a fixed, if controllable, electorate. While the votes of east-side blacks and west-side Hispanics often required a direct financial investment, the paying of these residents' poll taxes insured their willingness to vote as directed. Middle- and upper-class whites living in northern suburban tracts were less predictable. They represented a threat to the existence of the city machine and its supporters rather than an opportunity for community growth and expansion. As a result, local power brokers did not annex these peripheral neighborhoods, which would have been a financial burden due to their need for expensive capital improvements and infrastructure, and which would have increased the possibility of political instability due to their voting power. Instead, and for a time, local pols provided water lines and police protection to Olmos Park without cost. They provided these benefits knowing that they would not have to face the potential complications that this elite neighborhood filled with reform-minded voters might have posed. For its part, Olmos Park could maintain an image of independence from San Antonio even as it accepted key city services. This unusual relationship lasted until the late 1930s, when the city threatened to annex Olmos Park. Its residents immediately voted to incorporate, as did several other north side subdivisions, organizing to maintain their racial, class, and ethnic homogeneity.[5]

Yet even as the 1920s elite withdrew to shady suburban bastions, some of those people left behind benefitted from a boom in the construction of green space. This surprising development is also dependent on the odd political dynamic that sustained Olmos Park in its first two decades. The city's elected commissioners discovered that park creation secured votes and won elections, a calculation that depended in turn on the city's racial and ethnic electoral mix and a set of fiscal decisions that could keep commissioners in office.

c. 1910 postcard of Koehler Park. *Courtesy of the Edwards Aquifer Website by Gregg Eckhardt (http://www.edwardsaquifer.net/).*

The city's first open spaces—its parks and plazas—had been the gift of eighteenth-century Spanish planners. Another gift, Brackenridge Park (1899), named for its donor George Brackenridge, the city's most important entrepreneur of the late nineteenth and early twentieth centuries, became the 343-acre gem in otherwise paltry inventory of parks accessible to the public. Until the 1910s, housing developers were another source of new parklands, believing that their donations to the city enhanced the value of their real-estate schemes. This pattern shifted radically in the 1920s.[6]

During that decade, San Antonio sold a series of bonds that bought nearly 1,100 acres of land, more than doubling the size of the local park system. This transformation depended on an earlier change in the structure of local governance. In 1914, San Antonio had adopted the commission form of government, and one of its five members was the parks commissioner, giving this resource a motivated voice within the governing body. As a result, parks (and their elected representative) received a share of the funds that the bond issues would generate.[7] Citizens, for example, voted on the initial offering in 1919, which authorized the spending of $200,000 for parks, an initiative that won by a four-to-one margin. In 1923, the electorate agreed to the selling of $100,000 worth of bonds for parks. Three years later, the figure rose to $150,000. In 1928 another bond for parks totaling $650,000 won favor; and the tab for the 1930 bond, the last successful parks bond until the early 1960s, soared to $700,000.

The cumulative effect of those electoral victories, when combined with a quirk in San Antonio politics, was instrumental in this decade-long run of park-building. The bonds' promotors, for example, sold these bond elections to the city's working class by touting new parks as "the poor man's estate . . . vital to [the] health and well-being of the laboring people."[8] These citizens were also told that the "big property holder bears the brunt" of these new taxes, an appeal to class politics reflected in the geographical setting of the bond-funded parks.[9] Strikingly, five of the seven new sites were concentrated on the east side, home to much of the city's black population, with only one major park development located on the north side, largely populated by middle- and upper-income whites. That particular park, Olmos Basin, primarily served flood control and storm-water retention purposes. The city's west side gained but a single park in this period, and this, too, was designated for blacks only. At a time when racial segregation was the norm in San Antonio, the political process provided an unusual level of public benefit to the city's small black population.[10]

An answer for this conundrum lies in the fact that black votes played a critical role in passing every bond in the 1920s.[11] Not surprisingly, then, when the city commissioners extolled the virtues of parks for the laboring classes, they had a very loyal constituency in mind.[12] The tight relationship between African American voters and machine politicians was the consequence of east-sider Charles Bellinger's shrewd understanding of commission politics and the impact of bloc voting on its actions. The commissioners needed reliable votes, and Bellinger, working behind the scenes with the leaders of local African American churches, delivered them. "Negro voting and Bellinger's rule were tolerated by a corrupt city machine only because a large Negro vote, swung in the right direction, gave absolute assurance that the machine candidates would be elected," a contemporary journalist observed. The payoff was just as assured. "Through the power of his organized votes, [Bellinger] forced city officials to give special attention to the Negro quarter," so much so that the east side "is comparatively well paved; has adequate light, water, and sewer service; and enjoys numerous public schools, parks and playgrounds, fire and police stations, a public library and a public auditorium."[13]

The Great Depression, which in 1929 began to upend the national and local economy, would disrupt this reciprocal politics of exchange. So would Bellinger's death in 1937. Yet it is noteworthy that in formerly park-starved San Antonio, it was the much-reviled political machine, along with its key ally, Charles Bellinger, that manufactured the wherewithal and will to generate an array of public amenities that no previous form of government had provided. City parks, these professional politicians recognized, made for very good politics.

9
Deals Old and New

For a city whose economy depended in good part on military spending and tourism, the Great Depression was an unmitigated disaster. Although President Herbert Hoover increased federal spending in hopes of stimulating state and local governments to do the same (they did not) and urged voluntary organizations and private philanthropy to directly aid the unemployed and impoverished (they tried), he also he also signed the Smoot-Hawley Tariff (1930), which slashed international trade and further depressed further the global economy, and legislation that halted immigration, matter of considerable concern to cities like San Antonio with extensive trade relations with, and the flow of labor from, Mexico and the rest of Latin America. Alamo City merchants and government contractors also took careful notice of the president's 1929 assertion that "The American people should understand that current expenditure on strictly military activities of the Army and Navy constitutes the largest military budget of any nation in the world today and at a time when there is less real danger of extensive disturbance to peace than at any time in more than half a century." He concluded that the necessity of rebalancing the budget would depend "in large degree in our ability to economize on the military and naval expenditure and still maintain adequate defense"—a conclusion that Franklin Roosevelt pursued even more rigorously than his predecessor.[1]

Local entrepreneurs and commercial agents—and their employees—whose primary work was with the local military bases, were right to be anxious about the long-term viability of their business models. Civilian defense workers also received pink slips. At the same time, tourists stopped coming to town. Would-be travelers kept their disposable income at home, and the local service industry that had grown up around providing transit, entertainment, food, and lodging to people once eager to visit the city's legendary sites suffered. Those who hit the road, notably journalists and photographers assigned to record the nation's travails, flocked to San Antonio precisely because it was so hard hit.

They discovered that its poor, regardless of race or ethnicity, lived in squalor. That included a large number of San Antonio's 230,000 residents, particularly those Mexican migrants who had moved to the city in the 1910s and 1920s to escape the revolution and its aftermath south of the border. These new migrants jammed into the crowded west side and found that they did not have access to running water or indoor sanitation, that their streets were unpaved, and that their housing was substandard. Women, especially ethnic minorities, bore the brunt of this downturn, because their low-wage jobs in the garment, nut shelling, and other light industries evaporated. In keeping with Hoover-era anti-relief strategies that dominated federal, state and local responses to the depression between 1929 and 1933, San Antonio politicians resisted pleas to provide food, clothing and shelter for the community's most destitute. Yet they were also responding to the ominous spike in tax delinquency, which led city and county governments, as well as school districts, to cut staff and salaries drastically, adding to the community's woes. Resolving these intertwined issues in a social geography badly divided along lines of class, race, ethnicity, and gender would prove almost insurmountable, leading historian Julia Kirk Blackwelder to conclude: "No major city in the United States fought the Depression with fewer weapons than did San Antonio."[2]

To counter this grim reality required new political actors wielding innovative tools. One of those who helped changed the dynamic was President Franklin D. Roosevelt, whose inauguration in March 1933 and the fast-paced set of reforms his new administration launched during its first one hundred days signaled a marked shift in the federal approach to economic collapse and systemic poverty. Picking up where Herbert Hoover had left off, Roosevelt adopted some of the key principles associated with Keynesian economics, including the idea that governments can and should actively intervene via monetary and public policies to mitigate economic disruptions and social distress. This laid the basis for Roosevelt's active promotion of legislation to offer social security, set minimum wages, and establish labor standards. Over his first two terms in office, Roosevelt's administration advocated a slew of other projects designed to create jobs, provide relief, construct homes, and advance the commonweal. He and the Democratic Party-controlled Congress built the foundation for the welfare state and a web of sustaining safety nets that defined the national interest for the rest of the twentieth century.

Cities such as San Antonio were the logical recipients of federal aid—indeed, their political leaders, usually but not exclusively Democrats, were the Roosevelt administration's preferred recipients of funding.[3] Securing those dollars, however, required representatives in Washington willing to channel them to their hometowns. Rep. Maury Maverick embraced this role, after winning a congressional seat in 1934. A third generation San Antonian, Maverick's family had been engaged in the city's public affairs since the early nineteenth century; for him, public service was second nature. A member of the reform-minded, anti-machine San Antonio Citizens League, Maverick used this group as a platform to become Bexar County's tax collector, and from this office fought to increase relief efforts. He then campaigned for Congress successfully as a liberal Democrat in 1934, quickly becoming a national champion of the poor and dispossessed. Over the next two years, Maverick urged the White House to underwrite slum clearance and public housing, pressed the State of Texas to authorize a housing authority in San Antonio so it could act as the conduit for Public Works Administration (PWA) dollars, and worked in collaboration with a coalition of housing advocates, among them the San Antonio chapter of the League of United Latin American Citizens (LULAC). Although these earnest efforts did not produce "a single building in San Antonio," writes historian Robert B. Fairbanks, they "awakened a formerly apathetic city to its severe housing problem and educated citizens about how the slums threatened the physical and social health of the entire city."[4]

Father Carmelo Tranchese was as pivotal as Maverick to this educational enterprise. A Jesuit priest, in 1932 his order had assigned him to Our Lady of Guadalupe Church in the heart of the west-side barrio, which the WPA described as "one of the most extensive slums to be found in any American city."[5] Tanchese quickly recognized the truth of that claim, as most of his parishioners earned

Maury Maverick, c. 1936–37. *Library of Congress Prints and Photographs, Division Washington, D.C.*

less than three dollars a week shelling pecans, bussing tables, or from other forms of unskilled labor. Most of their meager income went to rent, and afforded them little shelter beyond a dirt-floor shack lacking utilities or toilets. Lacking basic drainage, even light rainstorms swamped the neighborhood, flushing human and animal effluent into homes and down narrow alleys and leaving behind pools of stagnant water in which mosquito swarms bred. Such fetid conditions, which a visiting reporter denounced as the shame of Texas, were also appalling relative to the much healthier environment that the city's African Americans inhabited due to Charles Bellinger's clout with the local machine. While Bellinger's "people live in a fairly decent section of town," this journalist observed, "San Antonio's ninety thousand Mexicans, lacking a clever and powerful political leader, live in the most revolting squalor."[6]

Combating these conditions became Tranchese's mission. Working in alliance with grassroots groups like LULAC, and politicians like Maverick, he gathered data about his congregants' working conditions, substandard housing, and levels of disease, destitution, and debility. The figures were shocking: in late 1933, more than 48,000 San Antonians were on relief, roughly half of whom were unskilled and semiskilled Mexican Americans. More than 68 percent were "on relief in 1935, 68.5 percent were still unemployed in 1937, and 85 percent were still registered with employment services in 1940," according to historian Richard A. García.[7] With reams of such statistics and revelatory photographs in hand, the indefatigable priest buttonholed officials at the city hall, county courthouse, and state capitol. He dispatched detailed reports to the PWA and President Roosevelt. Political machinations and personal rivalries at all levels of government blocked Tranchese's initiatives until the early 1940s, when Alazan-Apache Courts, the city's first public housing were completed with money from the Roosevelt administration.[8] In the interim, Tranchese poured his energy into on-the-ground activism. For the unemployed, he helped launched the Catholic Relief Association; to enhance his congregants' sense of belonging, he organized the Guadalupe Community Center; for the many afflicted with a disease generally associated with the nineteenth century, he used his position on the county's tuberculosis association to bring their plight to light. No wonder a sympathetic national profile anointed Tranchese "The Rumpled Angel of the Slums."[9]

Few thought union organizer Emma Tenayuca was as angelic as Tranchese. Yet like Tranchese and Maury Maverick, her sometime collaborators, Tenayuca was a force of nature. In 1933, at sixteen years old, she was rolling cigars at the Finck Cigar Company for the princely sum of 20 cents an hour (assuming she hit the daily quota of 500; if not, deductions kicked in). When the owner refused to up their hourly pay to 30 cents, as Roosevelt's National Recovery Administration recommended, Tenayuca and the all-female, Spanish-speaking workforce went on strike (and struck again, after Finck failed to adhere to the negotiated settlement). One year later, Tenayuca helped launch the city's Unemployed Councils providing relief for those out of work and joined women striking against the reduced wages that local garment factories paid. In 1935, the eighteen-year-old activist was leading the communist-affiliated Trade Union Unity League's operations in San

Emma Tenayuca in front of a jail cell. *Creekmore Fath Papers, e_rap_0105, The Dolph Briscoe Center for American History, the University of Texas at Austin.*

Antonio, growing its membership to 3000 or so, and was leading the Workers Alliance local. The energy these organizations brought to labor negotiations across the city provoked a brutal backlash. City Hall unleashed the police on the strikers and, armed with axes and clubs, they destroyed the Workers Alliance headquarters. The savage attack sent shock waves through the city. At Temple Beth El, Rabbi Ephraim Frisch, a tough-minded liberal, compared the police's thuggish actions to dictatorial violence in Italy, Germany, and the Soviet Union. The raid, he argued, "would have earned the designation of 'swell' from the mouths of Mussolini, Hitler, or Stalin if it had been performed by the lieutenants of these arch enemies of democracy."[10] Frisch's principled stand cost him the support of the synagogue's board of directors, which sought to silence Frisch, and deepened the anti-Semitic enmity that the city's ultraconservative Catholic Archbishop, Arthur J. Drossaerts, felt for the Jewish leader.[11]

Drossaerts was an equal-opportunity foe. At the same time he denounced laborers seeking wage justice, he offered unstinting praise for the city police and county sheriff officers who used unfettered force to roust protesters. Perhaps taking its clues from the Catholic prelate, LULAC distanced itself from the worker's plight and pleas. The waves of labor agitation that rolled through San Antonio, notes Richard A. García, "increased the cultural and ideological differences in the Mexican community; the *ricos* deplored the strikes and the middle class avoided them."[12]

This divide grew more pronounced in 1938. With Tenayuca at the fore, thousands of pecan shellers walked off the job. As they had during earlier strikes in 1934 and 1935, they protested against their miniscule wages and the cramped and unventilated quarters in which they toiled. Once more, the police, whose chief declared there was no strike so there could be no legitimate picket lines, retaliated with nightsticks and tear gas. Once more, the local archbishop red-baited the strikers: "Our police force has had a hard task of it these past three weeks. They fought, not the downtrodden sufferers of an egotistical capitalistic system, but the dangerous leadership trying to make hay while the communistic sun was apparently appearing above our San Antonio horizon."[13] With the oft-jailed Tenayuca sidelined, competing labor groups tried to manage the strike, and the final negotiations brought a modest boost in wages. This success proved fleeting. The Southern Pecan Shelling Company, which once boasted more than 10,000 employees, immediately mechanized its operations. By 1940, only 800 workers remained on the company's rolls.[14]

Although on the margins of this particular tumult, Maury Maverick was at the center of those demanding greater social responsibility from community leaders who routinely ignored or disdained the plight of the city's large number of unemployed and underemployed. His allegiances in this regard, combined with his anti-machine political perspectives, lost him his congressional seat

Pecan shellers removing nuts from shells at a non-union plant in San Antonio. *Farm Security Administration, Office of War Information Photograph Collection, Library of Congress Prints and Photographs Division, Washington, D.C.*

in 1938. Undeterred, the next year he ran for mayor as part of the Fusion ticket—a collection of liberal activists pitting themselves against Mayor C. K. Quin's patronage-driven control of the levers of power. Maverick campaigned hard on the west side, and despite failing to gain traction among black voters, whose interests had long been aligned with Quin and his predecessors, Maverick gained enough support on the white north side to enter city hall as the second Maverick to hold that high office (his grandfather, Samuel, served as mayor for two separate terms in the nineteenth century). Within months of Maverick's inauguration, Emma Tenayuca applied for a permit to hold a Communist Party rally at Municipal Auditorium. A staunch defender of the First Amendment, Maverick signed off on it. "No amount of pressure will force me to violate my oath to uphold the Constitution," he thundered.

"The right of assembly, and the expression of unpopular belief, are the bases for the preservation of democracy."[15] A mob of 5,000 rioters disagreed. On the evening of August 26, 1939, just as the official meeting began indoors, they descended on the auditorium, and started throwing rocks and bricks through the auditorium's windows. After the police escorted Tenayuca and her compatriots out of the building, the mob overwhelmed the police barricades, rampaged through the building and hanged Maverick in effigy. This disturbing episode is one of the reasons why Maverick lost his bid for reelection four years later.[16]

Despite the social rifts, punishing politics, and ongoing immiseration of too many impoverished San Antonians, the tumultuous decade brought some civic improvements. Leading the charge was the San Antonio Conservation Society (SACS). Founded in 1924, it was led by Anglo women eager to restore Spanish-era landmarks, a focus codified in the organization's official seal that contained an image of Mission San José's tower. After wresting control of the local preservation movement from Adina de Zavala, a key figure in efforts to protect the Alamo and other Spanish-era sites, these well-connected women used the personal touch with leading businesses, city and county government, the archdiocese, and state and national representatives. Politically shrewd, they used their social networks to secure funds and legitimacy for their cause, were in close contact with preservationists in California and the East to learn from their successes, and adopted, where necessary, a gradual approach to effecting their ends. Worried in the late 1920s that Mission San José would collapse before the society received permission from the church and abutting landowners to restore the edifice, they started to "buy pieces of land—a few feet at a time—surrounding the mission and granary and save the buildings from threatened encroachments," one of its leaders remembered. "What I couldn't buy we just fenced, since nobody seemed to own it."[17] New Deal largesse, which came with an array of supportive legislation, accelerated this piecemeal strategy.

One of those legislative initiatives, the Historic American Buildings Survey (1933), channeled money through the National Park Service to conduct site analyses that incorporated architectural renderings. Mission San José was the first to benefit from this baseline assessments, work that then led to a more considered rebuilding of its outer walls and Mission Indian dwellings that had been built into the interior of that boundary structure. A new church roof and the reconstruction of a hitherto-unknown mill were part of the restoration process. In 1941, SACS convinced the legislature to create the San José Mission State Park, forerunner of the current San Antonio Missions National Historical Park, which the National Park Service has managed since 1975.

Had Congressman Maury Maverick prevailed, the state park would have been folded into the national system in 1930s, but his lobbying efforts came to naught. That was a rare defeat for the energetic representative. During his single term in office, Maverick helped push through the Historic Sites Act (1935), which authorized the Department of the Interior to place national parks, monuments, and sites under the purview of the National Park Service. The statute also announced that hitherto it would be "national policy to preserve for public use historic sites, buildings, and objects of national significance."[18] This new standard energized local preservationists and Mayor Maverick would employ it in his efforts to refurbish a neighborhood known as La Villita. Although the Park Service would rebuff his bid for the federal agency to manage the site, La Villita was of undeniable significance. It was the location of an early Indian settlement, home to soldiers attached to Mission San Antonio de Valero (whose ramshackle dwellings the 1819 flood destroyed), and lore had that it was here in September 1835 that General Martín Perfecto Cos surrendered to Texian rebels in the run up to the 1836 Battle of the Alamo. By century's end, this historic neighborhood of adobe and wood structures housed a mix of new European immigrants. But by the mid-1930s La Villita had lost its charm. To reclaim its former beauty and boost its touristic appeal—becoming "a symbol and monument," Maverick enthused, "to those simple people that made possible the great city that had grown up around it"—the mayor tapped a stream of funding from the National Youth Administration (NYA) to pay for La Villita's repairs and upgrades.[19]

Curving beneath La Villata, the San Antonio River also became a major construction site when Maverick channeled $450,000 from the WPA to build the much-discussed River Walk; a river-centered improvement district cost an additional $75,000. The plan, which architect Robert Hugman had devised in the late 1920s, called for

Robert H. H. Hugman, EdwardsAquifer.net. *Courtesy of the Edwards Aquifer Website by Gregg Eckhardt (http://www.edwardsaquifer.net/).*

stone pathways to wind along the river's edge, limestone walls to shore up its banks, and above a series of patios, shops, restaurants, and cafes. The whole was to be transformed into "a sunken garden of loveliness."[20] Hugman's imaginative meld of aesthetics and commerce ran afoul of budgetary limitations and the San Antonio Conservation Society's competing vision of the River Walk as a naturally landscaped, linear park. Some of Hugman's ideas were implemented, perhaps the most striking of which is the Arneson River Theater, completed in 1941. An open-air performance space, with the stage on one bank and seating on the other, the theater links La Villita above with the river below. Many of the rest of Hugman's ideas were shelved along with the architect himself, as the oversight committee pursued a cheaper course that emphasized the river's park-like features. Not until HemisFair in 1968 identified the capacity for hotels, restaurants, and other amenities to integrate with the River Walk's ambiance would Hugman's original insights gain fuller expression.[21]

Other New Deal projects also drew visitors from near and far. The Alamo was the beneficiary of money from the Federal Emergency Relief Act (1934–38), as well as WPA and NYA contributions, to demolish unrelated structures on site, and build stone archways and perimeter walls. Upriver at the San Antonio Zoo (1939), sited in an old quarry, federal monies employed local stonemasons to cut animal enclosures into the limestone walls. Next door, some of the same limestone went into the construction of Alamo Stadium (1940), partial funding for which came from the WPA. Its design

Arneson River Theater postcard. *Courtesy of the Edwards Aquifer Website by Gregg Eckhardt (http://www.edwardsaquifer.net/).*

work secured additional funding from the federal agency's Arts and Crafts division; this included four ceramic tile murals that local artists Henry Wedemeyer and Leonora Feiler designed, and that a team of sixty fired in kilns at the south-side studio of Ethel Wilson Harris (who also supervised the local WPA office).[22] Even air travel received a boost from the WPA. Stinson Field, sited just to the west of the San Antonio River on Mission Road, lying between Mission San José to the north and Mission San Juan Capistrano to the south, gained a new terminal. Together, this slew of projects, and their physical locations, reveal a pattern of spending that bolstered a particular segment of the urban economy. They all are located on a north-south line running parallel with the San Antonio River. With the exception of the stadium, these developments sit within the river's floodplain, so any investment in them was predicated on the existence of the Olmos Dam and its flood-control protections. Each was also a consequence of their designers' and funders' ambitions to boost tourism in this historically poor city. Decades later, as tourism increasingly paid the city's bills, the dilemma that this low-wage sector posed for the restaurant workers, hotel staff, and parking-lot attendants it employed became much clearer than it did to those who constructed its foundations in the midst of the Great Depression.[23] "Federal money and a promising determination of Anglo, black, and Hispanic leaders to rescue San Antonio from its ignominious Depression reputation have improved living standards," Julia Blackwelder noted in the late 1990s. "Nevertheless, thousands of present-day San Antonians continue to know firsthand the perils of East Side or West Side life as it was in the 1930s."[24]

10

The Economy of War

World War II profoundly changed San Antonio. Yet for a city long accustomed to war and violence, it was not fully prepared for the demands, opportunities, and challenges that the second global conflict in a generation would generate. Mayor C. K. Quin signaled as much in the immediate aftermath of the Japanese attack on Pearl Harbor on December 7, 1941. He announced that in ten days, San Antonio would test its air raid system. Not that it had one, which was partly his point. After city council hastily drafted a blackout ordinance, Quin, knowing that there was no mechanism by which to signal lights-off or lights-on, improvised, asking factories and rail yards as well as the police and fire departments to blow their whistles and sirens at 9:00 p.m. on December 17. Businesses, hotels and shops, apartments and homes, and city and county offices were to cut their power; those requiring electricity painted their windows black. At the appointed hour, San Antonio went nearly dark. With an estimated 15,000 volunteers and wardens patrolling the streets, and a local radio station providing breathless live coverage, Quin and a host of civil defense and military officers crowded on the roof of the twenty-one-story Milam Building to observe as U.S. Army Air Corps pilots flew overhead, dropping flares to simulate an aerial attack. Although not hitch-free—a rash of broken arms resulted as people tripped down stairs or stumbled over curbs, and vigilantes hurled rocks at well-lit houses and stores—the hastily arranged but mostly effective spectacle put San Antonio on high alert. The war was on.[1]

San Antonians had not been caught off-guard, exactly. Since September 1931, when Japan invaded Manchuria, the first step in its decade-long, relentless assault on China and then Southeast

Jefferson Declaration

VOL. XI — SAN ANTONIO, TEXAS, THURSDAY, DECEMBER 18, 1941 — NO. 7

School Leaps to Defense of Country After Japanese Attack

Monticello Wins All-Texas Award

Annual Staff at Texas High School Press Association Meet at Denton

Masque & Gavel Gives Christmas Assembly

By Ray Keeler

Advisories Cheer 60 Families

Helpers Find Artists, Harvard Graduates Needing Christmas Aid

By Jane Allen Gillespie

Seniors Attend Kelly Field

Graduation Exercises Offer Annual Chance To Get Defense Snaps

Senators Win Sales

Members Lead Annual Tilt for Ninth Time

Nine Belong To Scouts

Defense Girls to Aid In Blackouts, Air Raids

University Professor Visits Art Class

LATIN CLUB IS FIRST TO PRESENT DEFENSE SAVINGS BOND TO SCHOOL

Left to Right: Raymond Heskin, Beverly Mae Jordan, Mr. T. Guy Rogers, Crickett Cannon.

Miss Oliphint Recalls Teaching Days in Japan

Skaggs Wins $1 Prize For Best Description Of Jeff's Poinsettias

THE POINSETTIAS

Hayes, Row, Skaggs Win in Poetry Meet

All Faculty Joins Red Cross — Bolton

Boys Enlist, Girls Take First Aid

Faculty Teaches What To Do in Air-raids; Clubs Buy Bonds

Miss Wolf Records Roosevelt's Address

Bob Roberts, '41, Tells How San Diego Marines Stand By

BOB ROBERTS

BOB ROBERTS

Orchestra Will Play Before School at Lunch On Last Two School Days

The *Jefferson Declaration*, the newspaper of Thomas Jefferson High School, tells of the school's efforts at the time of the United States' entry into World War II. *Frank Rosengren, Jr. (Frank Duane) Papers, UTSA Special Collections.*

Asia, it was difficult to miss the looming possibility that the conflict might sweep across the Pacific. It was not impossible to recognize, either, that a second, Atlantic-facing front might explode, and it did in September 1939. That is when the German army stormed into Poland, and eight months later crashed through the Netherlands and Belgium before racing into France. However distant these battles appeared to be from the vantage point of South Texas, and however much isolationist Texans and their compatriots across the nation hoped to avoid getting entangled in them, the Alamo City was among a handful of communities that early on experienced the benefits that war-preparedness could provide.

That was one of the lessons the city had learned during the 1920s. Even though once-busy Fort Sam Houston and Kelly and Brooks Fields seemed deserted as their functions and funding diminished rapidly at the end of World War I, soldiers and pilots continued to train at these bases. The city's leadership took notice. "Once accustomed to a steady diet of federal funds," historian David R. Johnson has noted, they "hungered for more."[2] And so did the U.S. Army Air Corps, which by the mid-1920s had determined it needed to increase its pilot-training capacity and to develop a more robust system of aircraft supply and maintenance. In 1927, the U.S. Army secured nearly $1.3 million from an otherwise tight-fisted Congress to build Randolph Field, with San Antonio scurrying to locate the necessary $550,000 to purchase a 2,300-acre site to the northeast of the city.[3] Construction dollars and military salaries flowed through the local economy, and grew with additional investments in and upgrades to Kelly and Brooks Fields along with the hiring of more civilian staff at the four installations. The link between this initial buildup and the war clouds looming on the horizon was evident when some pilots who had earned their wings over Randolph, the "West Point of the Air," voluntarily took to the skies against the Germans during the Battle of Britain (1940–41). Others enlisted in the Chinese Air Force as members of the First American Voluntary Group (1941–42), known also as the Flying Tigers, to harass the invading Japanese. On the ground in San Antonio, civic and political leaders were equally inspired, if differently focused. By the 1930s, many of them "had become committed to the idea that the military was a linchpin in the area's economy."[4]

Their realization dovetailed with the Roosevelt Administration's

Randolph Field. *Courtesy Tarrant County College District Archives, Fort Worth, Texas.*

accelerated fiscal commitments to defense preparedness. Beginning in 1936, in the run up to that year's presidential election, FDR began to funnel Works Progress Administration dollars away from boosting strictly civilian employment to construction work on military facilities. Kelly Field hired local labor to hammer in new barracks, officers' quarters, classrooms, and maintenance hangers; others poured concrete for an expanding network of runways, ramps, and aprons. In 1939, Kelly's civilian staff stood at 913; one year later it had zoomed to 20,000. More found employment doing similar work at Fort Sam Houston as wells as Brooks and Randolph Fields; the latter base was also expanding physically to absorb the hundreds of new recruits there receiving advanced flight training.[5] Even Stinson Field, the municipal airport on the south side, joined the pre-war effort. In 1940, Mayor Maury Maverick led a delegation of local officials and boosters to Washington, D.C., to lobby Congress for money to construct a new airport on the north side; the legislature concurred as part of its wider ambition to underwrite the development of upwards of 250 new airports across the country "as a matter of national defense." San Antonio's pitch contained a sweetener—it offered the recently rehabbed Stinson Field, located between to Brooks and Kelly Fields, to the Air Corps as an additional training facility. This proposition may account for the federal government's willingness to increase its contribution to the new city airport from an initial 20 percent of the final costs to 80 percent.[6] By December 1940, a year before overt hostilities between the United States, Germany, and Japan erupted, San Antonio not only had a keen appreciation that war that would be good for business, but that conflict was soon to come.

When war became a reality, San Antonio was overwhelmed by the flood of people who surged into the city. Hundreds of thousands of men and women were trained at San Antonio's major bases and on more than a dozen satellite facilities—one estimate puts the total number of trainees at more than one million. Their collective presence on the streets, in the stores and shops, along the River Walk, parks, and other open spaces, created a bustling downtown and filled cash registers. Those in the armed forces also dropped ready money in local bars, saloons, and gambling emporia. Movie theaters did a brisk business, as did the legendary red light district, just west of City Hall, and another that surrounded Sunset Station on the near east side. These latter environs kept the Military Police busy. The 1941 May Act had charged it and the FBI with preventing officers and enlistees from entering facilities perceived to be of ill repute, those identified for their "lewdness, assignation, or prostitution." The Office of the Provost Marshal posted lists of the off-limit establishments and neighborhoods, a publication that could have the opposite effect. This was especially true for gay and lesbian enlistees: "All a GI or WAC need do is read the list, understand the codes, and head out for a night of same-sex recreation. Ironically, the military imperative to regulate deviance facilitated the very behaviors such regulations were designed to stamp out."[7]

The military buildup coincided with a swift rise in San Antonio's civilian population. In 1940, it was home to 253,854; five years later, at war's end, it contained nearly 400,000 residents. Many of these new San Antonians had migrated from rural communities in Central and South Texas; they were drawn to the high wages and good benefits that that came with civilian defense positions. With the 1941 establishment of Lackland Field, which was carved out of Kelly and augmented with city-donated acreage, even more opportunities became available. Doctors, nurses, and other health-care workers were in great demand at Brook General Hospital located on Fort Sam Houston. The estimated range of federal expenditures in the Alamo City runs from one to several billion dollars, an infusion that powered a stunning economic turnaround. The previous twelve years had seen the collapse of the local economy; yet, with the declaration of war in December 1941, San Antonio, one of the nation's poorest cities, experienced a level of job creation that profoundly altered its people's lives.

During the Depression, for instance, Federal Emergency Relief Administration and Works Progress Administration projects had taught their employees new skills but even more importantly had "provided the margin of survival for many women and their families, though there was always more demand than either agency could begin to meet." That gap between the amount of available work and the number of people looking for employment was magnified in a troubling income statistic from 1940. Despite "the commencement in the late 1930s of the military buildup that brought jobs and consumer dollars to San Antonio," historian Julia Blackwelder notes, "San Antonio had the lowest median wage" of all American cities with populations of 200,000 or more.[8] By 1950, that was no longer true, a direct result of wartime spending that carried over into the post-war years, such that one in every three San Antonians worked on the local bases or in a supportive capacity off-site.

The distribution of these new jobs was skewed, reflecting the city's pervasive racial, ethnic, and gender divisions.[9] Access to the new public housing developments that cleared away acres and acres of dilapidated and pestilential shacks in the city's most distressed neighborhoods was similarly divided. Alazan and Apache Courts (1940–41) were reserved for Hispanics on the west side, Lincoln Heights and Wheatley Courts for blacks on the west and east sides (1940–42), and south-side whites moved into Victoria Courts (1940–42). But the fact was that there was civilian-defense work, with health and retirement benefits to go along with well-constructed housing, with running water and electricity, which was now available at low cost. That the city's Anglo elite had begun to recognize that the community's historic poverty damaged the health

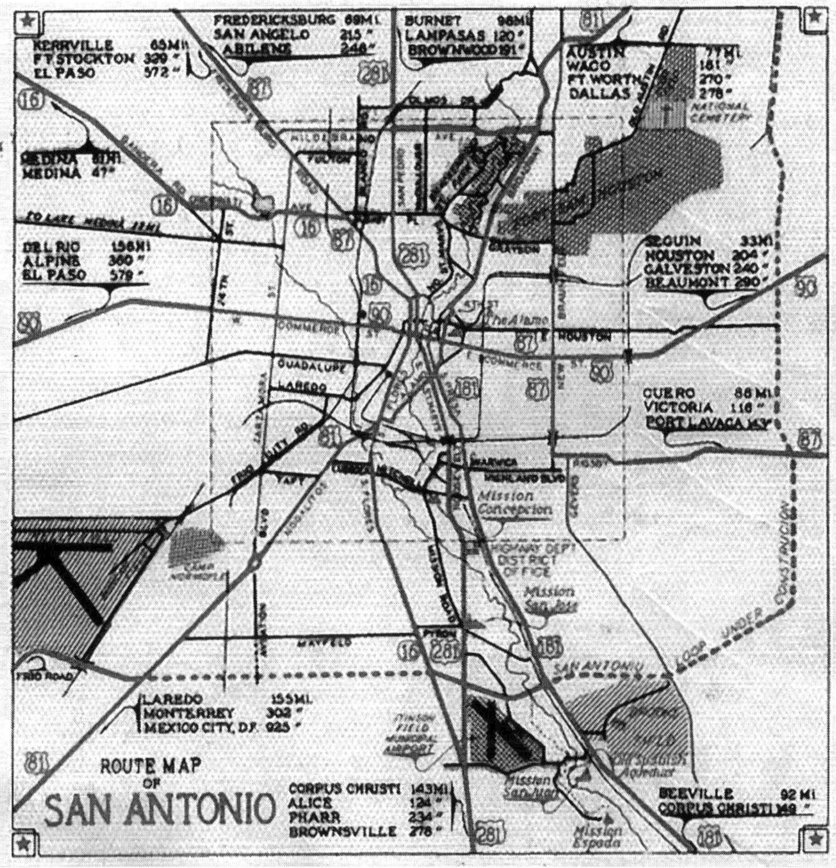

Map of San Antonio, 1941. From the *Texas Highway Map* (fall 1941), issued by the Texas Highway Department. *University of Texas Libraries.*

and wellbeing of the entire metropolis was a shift of immeasurable importance. So Mayor Maverick asserted in June 1940 at the dedication of the Alazan Courts, using the first person plural to herald the site's collective, revitalizing power: "This is one of the happiest days in our lives. I believe this is the most important day that San Antonio has witnessed."[10]

Another celebratory moment came, paradoxically, the day after the Japanese attack on Pearl Harbor, with the announcement that the peripatetic Trinity University, a Presbyterian institution that had been founded in 1869 in Tehuacana, Texas, and then in 1902 had relocated to Waxahachie, would move once again, merging with the Methodist-supported University of San Antonio. In 1942, the new institution took over the local institution's near west-side, Woodlawn Lake campus. The Chamber of Commerce, which helped underwrite the move because it wanted a strong Protestant college in a city with three Catholic universities, and yearned "with unutterable longing for a great university with a mighty football team" to fill nearby Alamo Stadium and compete for "pigskin glory," had more elevated plans.[11] By 1952, with a large number of veterans boosting Trinity's enrollment to 2,000 and munificent endowment funding from the city's Anglo elite, the university moved once again, this time to its 105-acre skyline campus on the north side, nestled within and above an old quarry, from which it gained a commanding view of downtown. The red-bricked university, whose mid-century modern design was the product of architect O'Neil Ford's guiding genius, shared this high ground with the all-white suburban bastions of Monte Vista, Olmos Park, and Alamo Heights, where many of the university's benefactors lived.[12]

The December 8, 1941, announcement of the city's investment in Trinity University's relocation was yoked to a much less happy relocation that same week. One of San Antonio's ethnic landmarks, the Japanese Tea Garden, had been built into a portion of the same abandoned quarry that the university and zoo would repurpose. But its name incited a xenophobic reaction in the immediate aftermath of Pearl Harbor, unlike its collaborative founding. The garden had been the shared dream of Parks Commissioner Ray Lambert and Eizo Jingu, a local artist and tea importer, whose late 1910s plans received a ringing endorsement from the *San Antonio Express* in 1918. "Mr. Lambert has been successful in securing the services of a woman, fresh from the land of Cherry Blossoms, who, together with her husband and two children, will make their home right on the park, where a cozy Jap [*sic*] house will be built for them, and tea will be served by this little lady dressed in her graceful flowing kimono, with all the fixin's."[13] When completed in the 1920s, the garden's pools, waterfalls, pagoda, and tearoom became a must-see attraction; the Jingu family lived on site and managed it until December 1941. With war declared, the city demanded that the Jingus leave; they refused, and in retaliation, the city shut off water and utilities as a prod to evict the family from the landscape they had stewarded for two decades. Once the Jingus left, the city then slapped a new name on it—the Chinese Tea Garden—and installed a local Chinese American family in the caretaker's quarters. It took more than forty years to right this wrong. One lever in this corrective process was that the Jingus two sons had promptly enlisted in the U.S. Army that December, and one of them, James, had fought in the much-vaunted 442nd Regimental Combat Team, a Japanese American unit. In the 1980s, Maury Maverick Jr., the former mayor's son, used their heroism to encourage the city to restore the site's original name. A city council representative, Van Archer, joined Maverick in this effort, and their arguments found favor with Henry Cisneros, San Antonio's first Hispanic mayor since Juan Seguín had held that post in 1841. To correct this particular injustice, and as part of his larger diplomatic play to woo Japanese industry to South Texas, Mayor Cisneros rededicated the Japanese Tea Gardens in 1984 with members of the Jingu family and Japanese government representatives in attendance.[14]

The Japanese Tea Garden, which was renamed the Chinese Tea Garden in 1941. *Courtesy of the Edwards Aquifer Website by Gregg Eckhardt (http://www.edwardsaquifer.net/).*

That Cisneros was in a position to act in this case depended in good measure on the activism that had driven returning Mexican American veterans—including members of Cisneros's extended family—to demand greater social equality and political rights for their communities. As had occurred after World War I, in 1945 Mexican Americans came back to San Antonio emboldened: they had earned their stripes abroad and were eager to advance their cause at home. Those who marched off to war from the west side, perhaps signing up with Rifle Company E, 2nd Battalion, 141st Infantry Regiment, 36th Infantry Division (Texas National Guard), the only all-Mexican American unit in the U.S. Army, found that the war had forged in them a Mexican American identity and a conviction that education and politics were key to "gaining knowledge and power." [15] In San Antonio, as in El Paso, Tucson, and Los Angeles, these men and women joined the G.I. Forum and LULAC to promote their concerns. They used the G.I. Bills of the 1940s and 1950s to secure advanced education and the skills that that training could provide, booming the enrollments at Incarnate Word College, St. Mary's University, and Our Lady of the Lake University. With access to federal housing mortgages, those who could afford to purchase homes did so. For others, San Antonio's new public housing served as a short-term steppingstone to the new subdivisions that extended out along the city's arterial highways, conveniently located near the military installations where an increasing number of Mexican Americans worked. "With the developing professional, service, and blue-collar job market, the Mexican-American population in San Antonio increased from 160,410 in 1950 to 243,627 in 1960," notes historian Richard A. García.[16] By the 1960s, this growing population accounted for more than 45 percent of the city's population.

A decade later, the economic drivers remained the same. Downtown businessman Morris Kallison's worries that if "we were to have peace tomorrow, we would find ourselves in the greatest depression we've ever known" proved to be unfounded.[17] The demobilization that he and others feared was swift, but also short. On March 5, 1946, former (and future) British Prime Minister Winston Churchill warned about the dire implications of a nuclear-armed Soviet Union and the Iron Curtain that it was dropping over Eastern Europe in a speech in Fulton, Missouri. To contain and confront this new global threat, one year later President Harry Truman committed U.S. financial and military resources to this new fight, and by 1948 was sending money and material to Turkey and Greece to shore up their defenses. The nation dispatched additional sums to prop up France's efforts to maintain its hold over its colonies in Southeast Asia. The Berlin Airlift of 1948, in which the U.S. airships carried a conveyor belt-like flow of supplies into the walled-off western sector of the embattled city, symbolized the Cold War realpolitik. It also highlighted San Antonio's contribution to this new, global struggle: to make certain that U.S. aircraft could fly non-stop into Berlin, Kelly Field's maintenance workers were called back on the line to work 24/7 repairing propellers, rebuilding engines, and fixing instruments.[18]

This was an important sign of what was to come. Defense spending sped up during the Cold War, with major spikes during the Korean and Vietnam Wars, such that San Antonio secured an estimated seven billion dollars between the 1950s and late 1970s, making the Alamo City a key center in the American Gunbelt.[19] This surge of capital turned Kelly Field into a massive industrial complex devoted to airplane repair and maintenance. It also transformed Brooks Air Force Base into Brooks Aerospace Medical Center, a site for cutting-edge research, and Lackland Air Force Base into the new armed service's Aviation Cadet Center. With few exceptions, all Air Force enlistees post-World War II began their careers in San Antonio. Randolph Field also morphed, becoming the service's go-to location for flight-instructor training. The numbers of recruits and officers who flowed through these sites is impressive. In January 1951, for instance, 55,000 recruits jammed onto Lackland, "with row upon row of squad tents, mess tents, and latrines on virtually every open space, including the main parade ground." The pressure was only slightly lower in the mid-1960s as recruitment ramped up in response to the escalation of hostilities in Vietnam, when Lackland "regularly handled recruit populations of 20,000 or more, while manned and equipped for 17,000 basic trainees."[20]

Of greater, long-term significance to San Antonio, though, were the increasing number of defense contractors—such as Southwest Research Institute (established 1947)—that provided the military with essential data analysis on fuels, engines, and lubricants or supplies, and others who supplied food, uniforms, and hardware.[21]

Critical too were the civilian employees needed to ensure the success of these bases' varied missions. Until the 1990s, when base closings and realignments began to diminish the local installations' economic importance and social impact, these installations provided an array of civil-service benefits, on-the-job training, and related advancement opportunities that solidified the prospects of countless San Antonians. The cross-generational impact was manifest as well in the expanded horizons now available to these employees' children. This rising generation would come of age in a San Antonio whose population in 1980 had topped one million and which counted Mexican Americans as the majority.

Before this demographic process had run its course, members of the postwar and highly self-conscious Mexican American middle class had begun to flex their political muscles. No one better represents this sector's emerging strength than Henry B. González. His parents and older siblings had migrated to the United States from Mexico, but Henry was born in the United States, and as an adolescent he recalled being the first of his family to stress that he was an American, "without prefix, suffix, apology or any other kind of modification."[22] Growing up on the city's west side, he graduated from the University of Texas at Austin and received his law degree from St. Mary's University Law School, close to his parents' home. During World War II, González served as a civilian cable and radio censor for military intelligence units. The ambitious young man entered city politics in 1953, winning a seat on the city council. He immediately challenged local segregation ordinances (with some success), and attacking discrimination in all its guises would become a mainstay of his activism at the state and national levels. In 1956, he was the first Mexican American elected to the Texas State Senate, and two years later he was the first Mexican American to run for governor. González then capped off his career in 1961, becoming the first Mexican American elected from Texas to the U.S. House of Representatives, a seat he held until he retired in 1998 (with his son, Charles, replacing him). González, like his contemporaries in LULAC, the Pan American Progressive Association, and other related organizations with a strong presence in San Antonio, embodied a set of cultural values that shaped his political perspectives: "ideological pragmatic, Americanist in its patriotism, and acutely conscious of its civic obligations to all Mexican Ameri-

Portrait of Henry B. González. *General Photograph Collection, UTSA Special Collections.*

cans regardless of class or status."[23] These principles, individually conceived and collectively expressed, empowered others to enter the public arena and secure positions of increasing importance in city and county government. It was on this generation's broad shoulders that political aspirants such as four-term mayor Henry Cisneros later would stand.

Nothing about this radical transformation was straightforward, fully realized, or unalloyed. Political challenges from the right and left periodically erupted. A series of pressing needs remained to be addressed: a more substantial K–12 educational system across the city, a more diverse economy demanding a highly skilled workforce, greater political accountability and inclusion, and more open space and rigorous environmental protections. Yet the ability to imagine that such needs existed depended on the powerful impact that the economy of war had on the lives and livelihoods of those residing in San Antonio. The city's belated recognition of this occurred in July 2017, when it officially patented its hitherto informal logo: Military City USA.[24]

11
Conflict, Consensus, and Change

HemisFair Arena was packed. The facility, located within the downtown convention-center complex and part of an urban redevelopment scheme constructed in advance of HemisFair '68, usually was jammed to the rafters with fans cheering their beloved San Antonio Spurs basketball team. On this day in 1984, however, those who crammed into the domed arena's narrow seating and endured its bad sightlines had come to pledge their allegiance to and celebrate the tenth anniversary of COPS. Formally known as Communities Organized for Public Service, in quick order the parish-based, neighborhood-focused COPS had become the single most powerful force in San Antonio's political life. Its ability to disrupt politics as usual, signaled by its appropriative presence in a building that resulted from the kind of closed-door decision-making that had long shut out Mexican Americans, was the central theme of Andy Sarabia's rousing speech to the assembled throng. Sarabia, the organization's founding president, reminded his rapt audience of the disdain they had confronted but ten years earlier. "City leaders said, 'Leave them alone. They're Mexicans. They can't organize.'" COPS made them regret those words, Sarabia thundered. "Today, we have power, we have our culture, we have our faith, we have our communities, we have our dignity, and we're still Mexicans. They feared the successful revolution from a government of the few by the few to a government of the people by the people and for the people. The significance is that the powerless do not have to stay powerless."[1]

Sarabia's tough-minded arguments must be set in their widest context. One contextualizing element concerns the (largely) white power elite, from which COPS's members claimed their newfound

A rally of the Communities Organized for Public Service (COPS), San Antonio. *Communities Organized for Public Service/Metro Alliance Records, UTSA Special Collections.*

power on their way to rearranging political relations in San Antonio. Those who had dominated the city since the 1950s, under the aegis of the Good Government League (GGL), had themselves wrested control from their predecessors. These self-styled proponents of good governance, disenchanted with the commission form of government that had shaped the city's politics since 1914, promised to clean up corruption and vote-buying; provide more and more-efficient health and educational services; and construct modern highways, hotels, and other urban amenities. Castigating the "machine" was as much a part of the GGL's pitch as its arguments in favor of modernity. The key to a more up-to-date San Antonio, its members believed, would be the adoption of a council-manager form of government consisting of an appointed city manager and an at-large-elected mayor and city council. This new system would allow these individuals to direct the community's affairs, impartially and with accountability.[2]

In postwar San Antonio, these claims resonated with many of the city's major business leaders as well as with suburban whites living on the north side and some reform-minded Hispanics and African Americans. Yet the transition was anything but placid. Supporters of the commission system—who counted among their allies a large number of downtown retail owners, in addition to voters on the east and west sides—fought to maintain their control. Elections in the 1940s and 1950s were bare-knuckle brawls as each side scrambled to build stable alliances and new constituencies. In 1948, hotel owner Jack White became the first mayor to favor a council-manager structure, but faced a set of commissioners hostile to its implementation. Making this the central issue of his reelection campaign two years later, White and a slate of likeminded candidates carried the day. Within months of their swearing in, a charter revision commission presented a new charter before voters, who passed it overwhelmingly. That should have ended the controversy, but in San Antonio politics there is no such thing as a done deal. Mayor White found that he enjoyed the relatively unfettered power the commission system provided and started to backpedal on the implementation of the voter-approved changeover. The GGL grew out of this fraught period, as reformers sought to create through this new entity a vehicle for implementing reforms and controlling city politics. It worked. Between 1955, when the council-manager governance structure finally went into place, and 1971, the GGL's secretive nominating committee, which pre-approved a slate of candidates for city council elections, won seventy-seven of the eighty-one seats in play; more than 75 percent of the successful officeholders were whites who lived on the north side.[3]

These elected officials distributed goods and benefits accordingly. Securing federal and state funding, they laid down highways and interstates that opened up large swaths of open ground on the north side for development. Other community resources, once slated to be built around the central core, were relocated. The county hospital and assorted clinics became anchors for a brand-new, north-side medical facility. Next to it rose a new campus of the University of Texas Health Sciences Center and farther north still was the site the council selected for the city's first public bachelor's degree-granting institution, the University of Texas at San Antonio. Around these and other economic generators grew new subdivisions, apartment buildings, and schools along with shopping malls and entertainment zones, all of which received publicly funded roads and sewers, water and other utilities. To insure that its incentivizing of this population shift (more accurately dubbed "white flight") did not undercut San Antonio's tax revenues, the GGL-controlled council used the state's flexible annexation laws to extend the city's political jurisdiction. In 1940, San Antonio was set within a 36 square-mile area, but by 1980 it had expanded to 262 square miles (a pattern that continued; as of 2017, the city encompassed 465 square miles, with additional annexations proposed).[4]

This bald favoritism for north-side development, COPS organizer Ernesto (Ernie) Cortés observed, "really sent signals of withdrawal of commitments to the older areas, withdrawal of commitment to the inner city."[5] He and others voiced the same critique of the GGL's reconceptualization of downtown as a haven for the tourist and convention crowd. To rebuild the hollowed-out central core that their public policies had created, the GGL chased additional federal and state dollars to bulldoze some of the city's oldest near-south-side neighborhoods and build structures needed first to house HemisFair '68 and that later would be converted to convention facilities. Attracting new hotels to house the fair's visitors proceeded apace. Palacio del Rio, a Hilton property directly across the street from the fair site and convention center, was the first to take advantage of the little-utilized River Walk. Refurbishing that linear park required new sources of money to extend it to a lagoon fronting the convention center, and then build shops, restaurants, and bars along its new network of paths. In 1973, a group of local investors, most of whom were ardent GGL supporters, purchased the Dallas Chaparrals of the American Basketball Association, renamed them the Spurs, and for the next twenty years the team played in HemisFair Arena, hard by the River Walk. The intensity of this initial wave of investment, and those succeeding it, created so much activity along the River Walk and its environs that by the 1990s it had supplanted those former staples of the tourist trade, the Alamo and the missions.[6]

The local black community was left out of this decision making process and failed to gain anything from it, argued Rev. Claude W. Black Jr. of the east side's Mount Zion Baptist Church. "Even though [HemisFair] abutted on the East Side, there was no spill-

A photograph showing the area that would be occupied by HemisFair '68. Houses, streets and other buildings that had to be demolished are faintly visible. The HemisFair site occupied 92.6 acres and was located on the edge of downtown San Antonio. *UTSA Special Collections.*

HemisFair '68 postcard. *Courtesy of the Edwards Aquifer Website by Gregg Eckhardt (http://www.edwardsaquifer.net/).*

over, jobs, or money. . . . On top of that they built HemisFair with no back door to the East Side. You had to go all the way around to Alamo Street to get into HemisFair. It was a message to us that this was not for the East Side." The city's budget conveyed the same message of exclusion: millions of dollars were poured into HemisFair-related infrastructure, an "allocation of funds spent in a way that left no concern for what was going on in the East Side community," Rev. Black asserted. "There were a whole lot of needs never addressed during that period."[7]

These manifold flows of capital to the suburban fringe and the downtown core, while they kept the GGL in power, also created the conditions that ultimately would shatter its hegemony. Early on, its promoters had promised the spread of city services across the map, but instead had concentrated their provision to the newer northside neighborhoods. The GGL deliberately neglected those sectors located on the east, south, and west, neighborhoods whose votes they did not look for or receive. Lacking flood control and suffering from poor drainage, neighborhoods on the west side in particular contained pockets desperate for running water, indoor plumbing, and sewer hookups; dysentery and related illnesses were rife. Those who lived on potholed streets, inhabited substandard housing, and sent their children to deficient schools offered COPS its defining issues and its shock troops.

Founded in 1973, COPS had the benefit of drawing on the tradition of local Latino activism that emerged following the two World Wars and the rise of a self-conscious Mexican American middle class. It also drew strength from 1960s youth activism focused on ending the poll tax and pursuing local funding from the War on Poverty and Model Cities programs. There was also the groundbreaking "work that people like Henry González and Albert Peña and Joe Bernal did," Ernie Cortés recalled, "in terms of trying to get people to think about things differently."[8] Cortés himself brought a distinct brand of organizing to the west side, which he had absorbed in the early 1970s while training at Saul Alinsky's Chicago-based Industrial Areas Foundation. One element of these tactics was to build the requisite base of support from the bottom up. Working through the close-knit network of parish advisory councils and tapping the energy of those who ran church-sponsored festivals, athletic leagues, and other activities, COPS strategists spoke with group after group to learn their felt needs. What had "pulling power," Cortés wanted to know? The answers were clear: floods and drainage, the failure of the city to provide basic utilities at affordable rates, street repairs and traffic control—"mundane issues that blighted their daily lives."[9] These interwoven concerns were immensely attractive to those who joined the grassroots organization, and with and for whom it spoke. They wanted to protect their families, hence the demand for a healthy physical environment, enhanced education, and the chance for the elders to remain, and the rising generation to settle, close to home. They wanted as well to preserve their communities *as* communities, as integral parts of their lived experience. "Neighborhood defines who they are," wrote one analyst; "neighborhood reminds them of where they have been; neighborhood is where they want to stay."[10] Theirs was a pragmatic, localized call to arms.[11]

They went public with their agenda after an August 1974 flood along Zarzamora Creek that rolled through an oft-inundated subdivision on the west side. As had happened in the past, homes were damaged, streets were scoured, garbage and other detritus piled up, leaving a rotten and stinking mess. COPS tapped the community's anger, called a meeting at a local high school and invited the city manager, Sam Granata, to explain why, once again, these Mexican American neighborhoods bore the disproportionate burden associated with flooding. Why had the city continued to refuse to invest in flood control on the west side? Ernie Cortés and the raucous crowd knew the answers. "San Antonio was run by a fairly well-knit oligarchy of well-off, very well-off people," Cortés later recalled. "All of whom, mostly whites and WASPS, lived in the northeast quadrant of the city and kind of dominated the politics and economics of the city," and therefore had little interest in investing "a whole lot of time and energy and attention or resources" on San Antonio's neglected neighborhoods, like those on the flood-prone west side.[12]

To fix these inequities required getting the city government and powerful institutions to recognize their complicity, a recognition that demanded confrontation. COPS activists rallied in department stores and banks, marched in the streets, and developed an innovative form of public hearing in which the audience held the power, not those seated on the platform. However confrontational, there was also a kind of decorum. "I sir'd him death," was Beatrice

Gallegos's recollection of how she spoke to Granata at the first of these events.[13] Others were blunter in testimonies that revealed the city's willful disregard of their plight. Asked if he would act immediately on the neighborhood's behalf, Granata acknowledged that as city manager he did not have final authority to resolve (or reverse) the historic problems that the crowd had enumerated. He promised to put the issue of flood-control on the agenda of the next city council meeting; he did so, and hundreds of COPS supporters showed up to demand action. Stunned council representatives listened to one speaker after another give voice to the west side's pent-up grievances against the GGL, which had done nothing to alleviate the environmental and social dilemmas that threatened the public's health and welfare. They also were caught off guard when COPS showed them the 1945 Master Plan containing commitments to channelize Zarzamora Creek and another document indicating that bond monies had been encumbered to underwrite its reconstruction but were never released for the project. The recently elected mayor, Charles Becker, gave city staff four hours to find the necessary funds and reconvened the council at midnight with news that the requisite dollars were available. The council had agreed to spend to underwrite a flood-channel project on the Zarzamora. COPS had won its first battle, but not its last.[14]

Within three months, the city council had agreed to a special bond election devoted to funding a series of flood-control initiatives on the west and south sides. This was the first of a remarkable $150 million it would funnel towards neighborhood improvements over the next three years. As impressive as these gains were, COPS was not simply interested in ameliorating the physical environment but in transforming the political landscape. That required attacking the source of the GGL's electoral power enshrined in the 1951 city charter that had mandated at-large elections for the mayor and city council. The Anglo north side, with its growing population, and whose electoral clout the council had increased with every annexation, voted overwhelming for the GGL. Because of this bias, COPS sued the city in federal court alleging that its annexation policy diluted minority voting power. The U.S. Department of Justice sided with COPS, arguing the city's charter was in violation of the Voting Rights Acts of 1964–65. Given the choice of voiding recent annexations or revising the city charter to create ten city-council districts, San Antonio capitulated. In early 1977, voters narrowly approved the revised charter, with the north side opposing and the west, south and east sides voting in favor. COPS had helped turn the "city's politics upside down."[15]

Unlike the GGL, which represented a "singular image of the common good," a good that it alone defined, COPS and its allies promoted a more diverse, particularistic conception of the city as comprising a series of distinct neighborhoods and interests. For them, the public interest was a "contradictory, dynamic, shifting notion expressing general and narrow values in the context of public conflict."[16] This vision was precisely what enabled a new and diverse coalition to emerge in hopes of slowing the hitherto unchecked growth of San Antonio across its northern rim. The key issue in this case was the protection of the Edwards Aquifer, a cavernous limestone geological structure that underlay the Edwards Plateau to the city's north and west. As the city sprawled out along its northern expressways, Interstate 10 and U.S. Highway 281, housing and commercial developments compromised the aquifer's recharge zones. Few paid attention until the early 1970s, when a massive housing project called San Antonio Ranch New Town received an $18 million loan guarantee from the U.S. Department of Housing and Urban Development (HUD). The project was to sprawl over 9,300 acres, and when completed house an estimated 88,000 people. Two new groups emerged, the Aquifer Protection Association, which activist Faye Sinkin directed, and Citizens for a Better Environment. They went to court seeking to halt construction due to its anticipated and detrimental impact on the aquifer. Although the lawsuit failed at every level in the courts, it won in the court of public opinion by joining with the Bexar County Commissioners, Representative Henry B. González, the Edwards Underground Water District, and other entities to combat the uncritical acceptance of growth for growth's sake.[17]

With Ranch Town came the imposition of a new kind of federal authority that the newly created U.S. Environmental Protection Agency deployed, courtesy of Representative González. Because activists had demonstrated that Ranch Town would compromise the complex mechanism governing the Edwards Aquifer's recharge, and because this groundwater was the city's "sole source" of potable water, González inserted an amendment to the Safe Water

Government Canyon State Natural Area. Source: https://tpwd.texas.gov/state-parks/government-canyon [Accessed Mar. 8, 2018].

Drinking Act of 1974. Targeted at "sole source" aquifers that supplied at least 50 percent of a community's water supply, it gave the EPA oversight of these resources; it could require, for example, that any project that received federal money—such as Ranch Town's $18 million from HUD—must include robust water-pollution protections. That amendment was a deal breaker locally, and it would have an equally dramatic impact across the country. One consequence of the failure to build Ranch Town was that these thousands of acres would remain open space, becoming Government Canyon State Natural Area in 2003. It continues to serve as an important catchment basin and recharge zone for the aquifer and educates the public about the interplay between the natural and built landscapes. The site also stands as a monument to the city's unusual dependence on this geological feature for its drinking water and the political struggle that preserved this hydrological system and the community's health.[18]

A second development—this time a regional shopping mall on 129 north-side acres that overlay the aquifer recharge zone—reaffirmed COPS's and local environmentalists' skepticism that the city would ever regulate growth. In October 1975, activists took to the streets and courts shortly after the city council signed off on the massive project. At stake, they asserted, was the storm water that would sheet off the mall's surrounding parking lots, flushing oil, gas, fiberglass from brake-pad linings, and heavy metals into the recharge sinkholes and fissures, befouling the community's sole source of water. The citizenry would be compelled to pay for the increased costs associated with purifying mall-generated pollution, an argument that fused social justice and economic inequities with anti-growth rhetoric and suspicion of city hall. COPS and the Aquifer Protection Association joined forces, launched a successful petition drive to place an anti-mall referendum before the electorate, and then in advance of the vote went house to house to gather support. The election was no contest: the grassroots opposition won by a nearly four-to-one margin. An in-house policy decision that would have furthered the ends of the growth machine had been "overturned by a democratic decision shaped in the crucible of open debate and controversy."[19]

Not all victories are complete, however, and that was the case in this particular instance: the mall developers sued, arguing that the state constitution did not allow voter referenda to overturn a city's zoning decision; the state supreme court concurred. An emboldened city, hoping to get ahead of the curve, pushed in two directions simultaneously. In May 1975, it hired an engineering firm to assess whether the proposed mall endangered the aquifer, and then months later expanded the analysis's range to include how much development would be possible without damaging the aquifer. While it awaited the results of the study, and in the aftermath of the May 1977 citywide elections—the first under the new charter that had created ten council-representative districts—the new council passed an ordinance banning all development over the recharge zone. Naturally, this act prompted another round of lawsuits, but the fact that those who formerly dominated all of San Antonio's policymaking now felt they had to litigate on behalf of their interests was a sign of their diminished power.

These legal struggles and political battles also set the stage for a new strategy for resolving the city's water needs. Conceding that northern growth would continue, and believing it essential to find a supplementary source of water, in 1979 Mayor Henry Cisneros pushed for and the council voted to authorize the city water board to procure the requisite permits and conduct the necessary environmental impact assessments for the construction of Applewhite Reservoir. Sited along a seven-mile stretch of the Medina River, flowing across the city's southwestern sector, the proposed reservoir became yet another hot-button issue. Opposition did not coalesce until the late 1980s, when the land had been purchased and the project had received regulatory approval. Opponents included those who had fought against Ranch Town and the north-side mall. Many of these critics argued that Applewhite's construction would have opened the floodgates to intense development over the Edwards Aquifer. Anti-tax forces, embodied in the Homeowner Taxpayer Association, also denounced the project as a "scheme to fleece ratepayers and provide construction firms and land developers with a windfall at taxpayers' expense."[20] Agitation and petition drives resulted in not one, but two, referenda on the reservoir. In May 1991, voters directed the city to cease construction on Applewhite. Seeking a way around this decision, Mayor Nelson Wolff, who had replaced Henry Cisneros after the charismatic leader had decided not to seek a fifth term, created the so-called 2050 Water Committee to assess all options, including reviving the reservoir initiative. Not a single opponent of Applewhite was included on the committee, whose membership was packed instead with engineers, developers, contractors, and public officials. Not surprisingly, they reported that the reservoir should be revitalized. The city council agreed. No wonder then, that local activists girded themselves for another brawl.

The opposition came armed with a new argument. In 1993, the state had created the Edwards Aquifer Authority (EAA) to manage this cavernous resource so critical to millions of Texans. It had done so because of a Sierra Club-funded lawsuit that alleged the U.S. Fish and Wildlife Service had failed to protect endangered and threatened species inhabiting the aquifer. Because the federal agency had not limited pumping of the aquifer, it had compromised the life chances of various reptiles, fish, and flora. The City of San Antonio had fought the suit and lost. Hoping to avoid ongoing litigation, the state legislature created the EAA, whose board would be elected from the counties that overlay the aquifer and whose powers to manage pumping were extensive. The EAA's regulatory clout spurred activists in San Antonio to contest Mayor Wolfe's and the 2050 Water Committee's decision to move ahead with the Applewhite Reservoir's construction. Wisely regulated and conservatively managed, they asserted, the aquifer would continue to fulfill San Antonio's water needs without Applewhite, and at a much lower

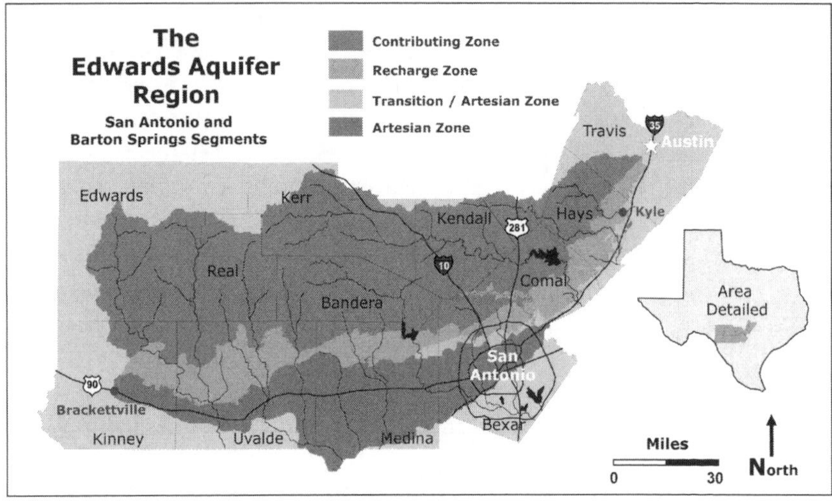

Map of the Edwards Aquifer. *Courtesy of the Edwards Aquifer Website by Gregg Eckhardt (http://www.edwardsaquifer.net/).*

cost. To the dismay of the city's powerbrokers, voters agreed: a second referendum on Applewhite met the same fate as the first, and by an even greater margin.[21]

These late twentieth century struggles over power and water—the two being parts of a whole—continued into the twenty-first. In the early 2000s, the development of PGA Village, an upscale golf resort that would be a joint project of the Lumberman's Investment Corporation and the Professional Golfers Association, ran into trouble winning community support. Securing a special taxing district from the state, the developers were required also to gain the city of San Antonio's approval before construction could proceed. The issue came before the city council in the waning days of Howard Peak's mayoralty (1997–2001), but the incoming mayor, Ed Garza (2001–05), stalled. In the interim, a new organization, the Smart Growth Coalition, emerged to pressure the council to vote against PGA Village. Among its partners was COPS, the Catholic Archdiocese, and social justice and environmental protection organizations. Over the next six years, in court and out, through city and county government deliberations, and at the state legislature, the project was vetted, denounced, refined, booted, rebooted, reconsidered and renounced, then finalized. In the end, PGA Village had morphed into the JW Marriott Hill Country Resort, built for $500 million, and opened with considerable fanfare in 2010. The majority of acreage remained as pervious surface and its irrigation system was closed-loop, meaning its waters would not seep into the aquifer, responding to key components of the environmentalists' demands. The resort's large workforce would be paid a living wage and receive full benefits (an essential point for COPS and the Church).[22]

These resolutions marked an important shift in the city's debates over water. So did another initiative that emerged at that time. Anxious to preserve as much of the imperiled aquifer's recharge and contributing zones as was still possible, in the late 1990s Fay Sinkin, dubbed the "Mother of Aquifer Protection" for her multi-decade-long activism to protect the city's groundwater resources, worked with a host of environmental and social justice groups to place Proposition 3 on the 2000 ballot. The proposition proposed

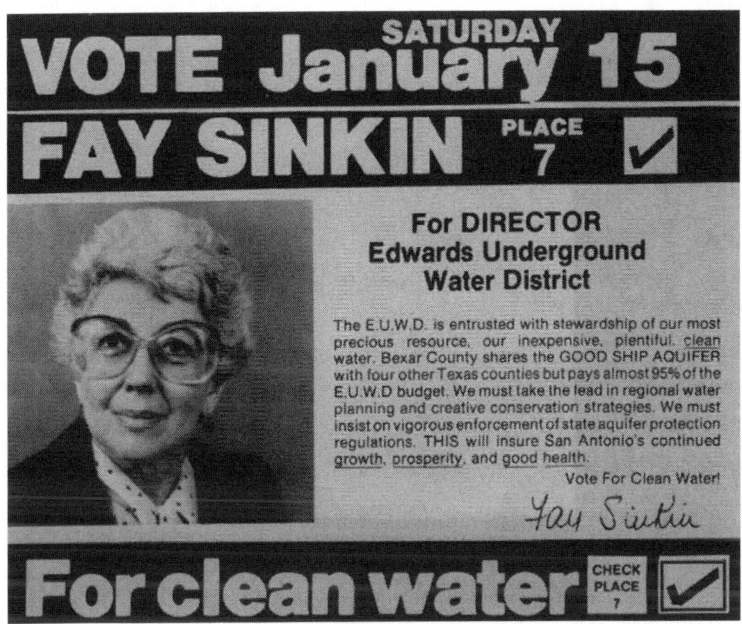

Fay Sinkin campaign flyer, 1983. *William and Fay Sinkin Papers, 1928–2008, MS 64, UTSA Libraries Special Collections.*

raising the city's sales tax by one-eighth of a cent, and the resulting funds would purchase recharge-features of the Edwards Aquifer. It was the only tax-based initiative to secure voter approval that year, and this landmark achievement enabled the city to acquire upwards of 6500 acres. Its success inspired voters to back similar measures in 2005, 2010, and 2015, and to date the city has been able to purchase more than 96,000 acres of sensitive watersheds, recharge structures, habitat, and open space.[23]

Sinkin, who died in 2009, would be the first to observe that this land-purchase process could not compete with developmental capital, which can—and has—outspent the city to power massive growth ever farther north from the downtown core. Yet, it is also true that the existence of these aquifer-protection propositions, like the concessions that activists gained through their negotiations with the golf-resort developers, would not have occurred had not COPS and its allies in the early 1970s stormed the civic arena. They demonstrated that growth and development were not givens. That the public good (and goods) were divisible and contestable. That the best way—the only way—to build a more habitable and just community was to fight for it.[24]

12
FUTURE SHINING?

Mayors like to leave telltale marks, enduring signs of their legacy. No one left a larger array of these signifiers than Henry Cisneros. A partial listing of those projects completed or started during his four terms in office includes: the San Antonio River Tunnels (1987–97), Vista Verde Project (1988), SeaWorld (1988), Rivercenter Mall (1988), The Alamodome (1993), Fiesta Texas (now Six Flags over Texas, 1992) and the Hyatt Hill Country Resort (1993). These and others, whether conceived, chatted up, or constructed, were part of Cisneros's concerted effort to remake the city of his birth and the office of mayor itself. Not all turned out as planned. Vista Verde, and, most notoriously, its anchor project, Fiesta Plaza Mall, failed to revitalize the near west side. Sea World and Fiesta Texas may have offered tourists additional reasons to remain longer in San Antonio, but as generators of secondary economic activity they were located so far out on the city's western and northern margins that it took decades for proposed housing and services to fill in the surrounding environs. As for the Alamodome, it has never lived up to the grand expectations its construction generated. That said, the pair of river tunnels bored under the downtown core have proved successful in sluicing floodwaters away from the downtown core. Rivercenter Mall, with a new extension of the River Walk pooling in front of it, and an associated Marriott Hotel slotted next door, has lured in locals and visitors alike.[1]

Whatever this complex record of wins and losses reveals and however self-congratulatory the boosterism that underlay these initiatives, it testifies as well to the game-changing nature of Cisneros's political vision. He pulled in tens of millions of dollars from the federal government and private investors, and matched those

Four-term San Antonio mayor Henry Cisneros, official portrait for U.S. Office of the Department of Housing and Urban Development. *Wikimedia Commons*. Source: https://commons.wikimedia.org/wiki/File:Henry_Cisneros_official_portrait.jpg [Accessed Jan. 19, 2018].

investments with a substantial commitment of local bond monies. The efforts energized San Antonio's sense of itself, no mean feat.[2]

Those who succeeded Cisneros in office have adopted some of his strategies, not least the urge to leave their physical mark on the built and natural landscape via signature projects. Lila Cockrell (1989–91), who entered politics via the League of Women Voters and the Good Government League, may have been the exception to this rule, having "depressurized the city after the frenzied Cisneros years," but her successors did not take their cues from her equanimity.[3] Nelson Wolff (1991–95) brought the Alamodome to completion, and thus owns a piece of its less-than-stellar legacy (and as county judge has been as active in bringing other major initiatives to completion, including the AT&T Center, the Toyota manufacturing plant, PGA Village, and extending the River Walk's reach, north

Photograph of a river barge as it enters the lagoon at River Center Mall in downtown San Antonio. *Evelyn Streng Slide Collection, Texas Lutheran University.*

and south).[4] In his single term, William Thornton (1995–97) started work on Port San Antonio, a proposed manufacturing and logistics center located on the site of the Kelly Air Force Base. Thornton and other hoped Port San Antonio might compensate for some of the jobs lost with the base's closing, a process with more fits than starts. Reflecting his rise from neighborhood activist to zoning commissioner and city council representative, Howard Peak (1997–2001) focused on initiatives with seemingly smaller footprints, such as a set of linear hike-and-bike greenways set within the floodplains of the community's major waterways (a system that in 2013 would be named for Peak).[5] Ed Garza (2001–05), an urban planner like Peak and Cisneros, developed City South, a New Urbanism live-work community that famed architects Andrés Duany and Elizabeth Plater-Zyberk initially designed. Part of Garza's push for a more balanced growth in a city top-heavy with north-side developments, City South was planned for that neglected region with the goal of creating new and affordable housing and well-paying work at the Toyota Tundra manufacturing plant and its suppliers.[6] Longtime federal judge Phil Hardberger (2005–09) built on his immediate predecessor's goals, and added his own twist. Even as he used tax abatements to lure new employment centers and educational opportunities, Hardberger also invested bond money to rehabilitate Main Plaza, a key hub in the downtown core since the eighteenth century; he also secured federal, state, and local funds to extend the River Walk to the south (the so-called Mission Reach) and to the north (the Museum Reach). These came to full fruition after his mayoralty, as did his efforts to counter San Antonio's dearth of open space on the autocentric north side. Hardberger dipped into local financing to purchase and develop Voelcker Park, which the city named for him after he left office. Not the retiring sort, in June 2017 Hardberger was a driving force behind a successful $850 million bond package that included money to construct a land bridge across Wurzbach Parkway to unite two sections of the park.[7] Before he left city hall to serve as President Barack Obama's Secretary of Housing and Urban Development, Julian Castro (2009–14) laid the foundations for what he dubbed "the Decade for Downtown." This ambitious scheme anticipated reversing the post-World War II rush to suburbia by incentivizing new residential and commercial development in and around the historic core, creating a more pedestrian friendly, millennial-attracting, and vibrant streetscape.[8] These were not the foci of Ivy Taylor (2014–17), the first African American to serve as mayor, and the first woman to hold that position since Lila Cockrell stepped down twenty-three years earlier. In a city in which upwards of 15 percent of the population live below the poverty line, her emphasis on job creation made sense. More controversial was her linking of cross-generational impoverishment with poor people "not being in a relationship with their Creator," a late-campaign comment that undercut her 2017 bid for a second term.[9]

Enumerating these mayoral initiatives is consistent with Henry Cisneros's profound impact on the job of the city's elected leader, turning the mayor into San Antonio's public face. Yet, this synoptic listing also elides the critical processes of negotiation and consultation with city council and the city manager's offices; with neighborhoods, civic, religious, and business leaders; and the larger political contexts—local, state, and national—that also shaped what any mayor could or could not achieve. Consider the local impact that the wind down of the Cold War brought with it, the crimping of the hose-like spray of federal dollars that had poured into defense-industry cities such as San Antonio. The Base Closure and Realignment Commission Act of 1990 (BRAC), which established a nine-member commission to transfer and dispose of military installations considered inessential, inefficient, or unnecessary for the nation's defense, was a project that brought sleepless nights to many San Antonians. Well aware that its five major bases were the single greatest contributor to its economy, community leaders fought against reductions of funding, civilian-defense work, and the reassignment of base missions. Their rearguard actions were not successful. In 1995, BRAC commissioners set the timetable for the closure of Kelly Air Force Base (AFB), with the consolidation of some of its workload to other Air Force depots; Lackland AFB absorbed some of the remaining missions. In August 2001, the Department of Defense formally transferred ownership of 1,900 acres of runways, hangers and other facilities to Port San Antonio, whose central task has been to redevelop the once-massive base. One year later, Brooks AFB was also decommissioned, and over the next fifteen years, the newly named Brooks-City Field slowly evolved into a south-side hub for medical, educational, and pharmaceutical research and training.[10]

Making these difficult economic transitions even more so was a radical change in the length of time an individual could serve on the city council. As mayor, Henry Cisneros was able to get so much done in part because he was in office so long. Equally impressive was the cumulative record of Councilwoman Maria Antonietta Berriozábal, the first Latina elected to the council. She became a force to be reckoned with because she had a thorough understanding of the city's political dynamics hard-won over her decade on the council.[11] But in 1991, after a series of high-profile scandals involving local politicians, and angered by what they perceived as an out-of-control, tax-and-spend local government, voters successfully amended the city charter to limit the mayor and council representatives to two terms of two years each, with a lifetime ban on subsequent service. This was the toughest set of restrictions of any major city in the United States and had been the brainchild of the conservative Homeowner Taxpayer Association (HTA). Its passage emboldened the HTA to demand, and in some cases, secure decreases in property taxes (already among the lowest in the state) and a reduction in city services. HTA also opposed spending taxpayer dollars on air-conditioning the classrooms in the San Antonio Independent School District and fluoridating public water. This constrained environment complicated the capacity of city hall to defend local military bases and advocate for long-term development of the local economy. The anti-tax fervor also troubled San Antonio's bond ratings across the 1990s, making it more expensive to borrow needed funds. Curiously, however, these various challenges to effective governance became one major reason why mayors in the post-Cisneros era pursued landmark projects as a quick way to encapsulate their short shelf life.

The greatest impact of the system of term limits was built into the short terms themselves: between 1989 and 2009, eight different mayors and more than fifty council representatives held office. Local politics had become a game of musical chairs: the first year in office, council members learned their job; year two, they ran for a second term; if reelected, then in year three, they governed; in year four, they either ran for mayor, sought a seat on the county commission, or looked for another line of work. This "churn of mayors and council members," wrote Nelson Wolff, the first mayor affected by the new restrictions, "resulted in inexperienced mayors and council members. It became clear that the voters had driven a stake into the heart of City Hall."[12]

This revolving door had multiple outcomes. Because of a weakened city council, stability lay with city staff, whose power increased as mayors and council members relied on them to guide them through the legislative processes. Stark disparities in salary reflected these differences in clout. Until 2015, council representatives received $20 per day per session, a figure set in 1951. The mayor received a meager salary of $4,040. By contrast, in 2015 City Manager Sheryl Sculley's base pay was more than $369,000. Realistically, only those who could afford to work without an income, or those who held a second, full-time job, could serve on the council or as mayor. That awkward situation further decreased the elected representatives' capacity to govern effectively. Yet voters rebuffed several attempts to loosen the strict term-limits law and thus recalibrate the tenor and balance of local politics. In fall 2008, however, Mayor Hardberger advocated, and voters accepted, a modest revision to the charter that granted council members and the mayor the possibility of two four-year terms. Seven years later, the electorate finally agreed to pay city council representatives and the mayor a decent wage.[13]

Another consequence of a weakened city council was that developers, who had long dominated the political arena, proved even more indispensable to first-time candidates and neophyte council representatives. Their campaign donations could make the difference between winning and losing, and that gave them an advantage and influence they might not otherwise have had. During his tenure, for example, Mayor Howard Peak diligently negotiated a new Urban Development Code intended to control runaway suburbanization and increase density, but developers and contractors, who served on the requisite subcommittees, weakened its regulations before final passage; they also girdled a tree-preservation ordinance that would have greened up the city. Even before the city's political representatives gained the potential to serve two four-year terms, developers had gone to the state legislature to handcuff city ordinances that might impinge on their construction projects. The so-called "vested rights" law prohibits Texas cities from imposing restrictions on a project once a developer has submitted any kind of plat or master plan, regardless of when construction begins (if it ever does). Whenever the city announces an increase in stormwater and sewer fees or a boost in oversight of the growth over the Edwards Aquifer, contractors, engineering firms, and land speculators immediately submit plats and other informal plans to secure their vested rights. "By almost every measure," the *San Antonio Express-News* reported in 2005, "the statute has been a boon to developers while hampering efforts by residents, community groups and officials to make San Antonio a better place to live."[14]

A final consequence of the city's controversial term limits was that Bexar County government officials began to fill the void in local political leadership. Its commissioners, some of whom had been termed-out from city council, and led by two strong county judges—Cindi Taylor Krier (1993–2001) and former mayor Nelson Wolff (1993–present)—shed their historic backseat role. In what he describes as a "shot across the mast of City Hall, signaling that it was about to emerge as a power player," Judge Wolff recounts how Krier outmaneuvered the inexperienced mayor and council to build the AT&T Center, the new home for the San Antonio Spurs, on *county* property. "After that victory, county government emerged from the city's shadow and became a respected entity."[15]

Amid these shifting dynamics in local politics, San Antonio continued to grow rapidly. In the 2010 census, the population stood at 1,327,538; six years later, it was estimated to be 1,492,510, an increase of more than 12 percent. Hispanics or Latinos constituted 63.2 percent of the population, whites 26.6, African American 6.9, and Asian Americans a little more than 2.[16] The growth of the city's physical expanse, which in previous decades had ballooned outward due to annexations that allowed San Antonio to claim to be the nation's seventh largest city and the state's second largest, slowed. In 2000, it managed 412 square miles, and in 2017 it controlled a touch more than 465 square miles. San Antonio's relative size shrinks when calculated as a metropolitan area; encompassing eight counties, in 2016 San Antonio was the nation's twenty-fifth largest metro area with a total population of 2.429 million. Hidden within that data is a measure of the city's sprawling extent. Like residents of other southwestern cities that grew largely in tandem with the automobile and high-speed expressways, those living in San Antonio who could afford to move away from the central core did so. Their outward flight to green-lawn suburbia kept density low even as it intensified daily commuting on the very freeways and highways critical to those subdivisions' construction. As the population boomed, traffic correspondingly spiked, such that by the late 1990s for the first time San Antonio began to exceed the federal EPA's national ambient air-quality standards. Although it did not reach "nonattainment" levels, which would have brought stricter oversight from the EPA, at various points in the 2000s the Alamo City had to submit voluntary plans to control its ozone levels. By 2015, San Antonio—whose economic growth was fueled in part by the fracking-generated, oil-and-gas development to its south—realized that emissions from the Eagle Ford play would break its claim of "being the only large U.S. city with a perfect record on federal Clean Air Act compliance."[17]

Other environmental worries clouded the city's future. Water long has been the community's most tenacious issue. To supply its growing population, since the 1950s city council after city council has tried to procure more water through reservoirs, cross-basin pipelines, and desalinization projects. The battles over these projects foreshadowed the one that would surface in reaction to the Vista Ridge pipeline initiative. In March 2014, the San Antonio Water System (SAWS) publicly rejected proposals to pipe water into the community from distant watersheds in favor of desalinization. In private, it actually was negotiating with Abengoa, S.A., a Spanish investment firm, to build a 142-mile-long pipeline northeast to Burleson County. Once completed, the "San Antonio hose" would sluice 16.3 billions of gallons of water each year to SAWS customers. The city's power elite met behind closed doors to develop the deal, and SAWS suppressed a report critical of the project that indicated it had a "high risk" of failure (and then produced a revised version that substantially downgraded this once-critical scientific finding). As protests in Burleson and Bexar Counties mounted, Abengoa's serious financial woes—which SAWS knew about in advance but which it did not share with the city council—led the water purveyor to cancel its contract with the firm. In 2016, Garney Construction took over the project. The gain in potable water supplies came at a cost. Councilman Ron Nirenberg, who would become mayor in 2017, argued that one result of SAWS's lack of transparency and accountability was a more distrustful council: "in terms of moving forward, it underscores the need for oversight," he told the *Express-News*. "We have to be doubly sure that council, as the voice of the citizens, has a role in providing proper oversight of major projects." Another was an angered public, who as a set of frustrated ratepayers once again would foot the bill with little voice in local deliberations. "I think it cost some relationships with the constituents and some relationships with the ratepayers," declared Calvin Finch, who had written the original, suppressed analysis of the project. "We spent all these years making San Antonio ratepayers sophisticated in terms of water resources, and then we attempt to block a report that was important to a major project. What did it accomplish? Trying to block it didn't accomplish anything. It cost trust and it cost relationships."[18]

This rush to capture white gold and the economic bonanza the pipeline's boosters believed it would generate was also predicated on the presumed need to procure as much water as possible in advance of population projections. By 2040, the San Antonio metropolitan area is expected to increase by more than 50 percent, topping 3.750 million people.[19] This projection emerged as an issue in the 2017 mayoralty campaign, with the eventual winner, Ron Nirenberg, declaring in a campaign brochure: "The 'next million' are coming and have high expectations for our City, just as we do. We welcome growth, but we do not deserve to sacrifice our own quality of life. We must ensure that the people who have called San Antonio home are provided the same comforts as those who are coming."[20]

Those who already arrived were remaking San Antonio for the better, *Forbes* assured its readers. The city's rapid growth in new work—"Since 2000, San Antonio has clocked 31.1% job growth, slightly behind Houston, but more than twice that of New York, and almost three times that of San Francisco and Los Angeles"—came coupled with a similar boom in work and people in Austin. This led the magazine to declare that the two Texas cities, tied together by Interstate 35, were "America's Next Great Metropolis." The southern anchor of this urbanizing region, long "considered a laggard, a somewhat sleepy Latino town with great food and tourist attractions and a slow pace of life," was in fact nothing of the kind. Economic and business data demonstrated the fast uptick in STEM-focused industries, financial services, and cyber-security—"not in hospitality, and low-end services, but in the upper echelon of employment." The future looked bright, *Forbes* predicted, especially because of "the massive in-migration into San Antonio. Long seen as a place dominated by people who grew up there, the metro area has become a magnet for new arrivals. Since 2010, its rate of net domestic in-migration trails only Austin among the major Texas cities. Significantly, the area's educated millennial population growth ranks in the top 10 of America's big cities, just about even with Austin, and well ahead of such touted "brain centers" as Boston, New York, and San Francisco."[21]

This forward-looking, even giddy prognostication, allowed local promoters to gloss over the corrosive impact that residual poverty, spatial segregation, and social inequality have had on the Alamo City. In 2016, the U.S. Census Bureau noted, more than 275,000 San Antonians lived below the poverty line, or roughly 20 percent; nationally, the figure stood at 13.5 percent. That aggregate number did not tell the whole story, for this level of distress can soar to 60 percent in those neighborhoods on the west, south, and east sides that historically had been the most impoverished. Of San Antonio's sixty-eight zip codes, thirty-one had poverty levels exceeding 20 percent. These enduring concentrations posed serious political issues that the *Forbes* report ignored. "You see a divide in San Antonio you don't see anywhere else," observed Steve Glickman, executive director of the Washington-based Economic Innovation Group, an observation contained in a major investigative report published in the *San Antonio Express-News*. "There are huge disparities between ZIP codes, making the city not just economically segregated, but geographically segregated, to a shocking degree." So distinct were the well-off from the down-and-out that it was as if they inhabited "two different countries."[22]

Embedded in these stark disparities were staggering educational inequities. In 2016, young Hispanics, the largest single group in the city's poverty statistics, attended local schools that received some of the worst ratings in Texas. This is a partial result of residual discrimination: San Antonio contains sixteen independent school districts that when they were established reinforced prevailing forms of racial and ethnic segregation. Inequitable school funding has added to these woes. Until the 1990s, local property taxes underwrote school budgets, enabling richer and whiter districts on the city's north side to spend exponentially more money per child than those on the west, east, and south sides. In 1993, after a series of lawsuits demonstrating this system's discriminatory impact, the Texas state legislature finally adopted a court-mandated equitable school-funding mechanism (that its opponents promptly dubbed the "Robin Hood Plan"). In time, this new formula may bolster the educational possibilities for those in the city's Latino-majority schools, but that process will take years to accomplish. In the meantime, the San Antonio, Edgewood, Southside, and South San Antonio Independent School Districts continued to record the city's highest dropout rates and the lowest graduation rates. The social ramifications of these inequities are disturbing: nearly 50 percent of those who did not graduate from high school live below the poverty line, an indication of systemic failure. To counter this troubling outcome, every modern mayor and a coalition of civic leaders, activists, and ordinary citizens have routinely called for school reform, knowing that providing a high-quality education for all children is key to San Antonio's ability to create greater economic opportunities and a more just society.[23]

Securing justice requires an enduring commitment to community, a profound sense of shared experience. If elite enclaves and wealthy school districts, for example, do not imagine that they inhabit the same city as those who are poor and poorly educated, then the sharp divides that cut along class, ethnic, and racial lines will continue to haunt the city's prospects across the twenty-first century. In the words of former councilwoman and long-time political activist Maria Berriozábal: "we still consider growth for growth's sake as 'economic development.' We have not stopped to analyze how this kind of growth—which typically provides material benefits to wealthier people and low-wage service jobs to working people, with indirect costs to all—impacts the overall health and well-being of our city. We have never assessed the long term consequences of our unsustainable growth patterns."[24]

A different division has complicated efforts to develop a more livable downtown framed around alternative forms of transportation. "There is simply no broad community buy-in for the notion that San Antonio *should* have a thriving urban core," one critic observed. "The concept is one of many that falls into the yawning divide that separates suburban and urban San Antonio."[25] To bridge these varied demarcations scored into the ground requires a compelling conception of the city *as* a city. One suggestion for how this might occur came from the late Trinity University anthropologist John Donahue, who wrote about San Antonio's longstanding and bruising battles over water. He once noted that this particular struggle would not be resolved until the contestants adopted "a common language with which to talk about the management of the aquifer." Such an inclusive language must grow out of a "moral community" dedicated to managing this vital resource as a commons, the construction of which also depended on giving everyone an equal voice and a "place at the table." The analogy works across the board: learning to speak as one is how the many will move ahead, a telling mark of how successful San Antonio will be in 2118, when it celebrates its quadricentennial.[26]

Restoration: An Afterword

San Antonio's eighteenth-century past received international acclaim on July 5, 2016, when the United Nations (UN) bestowed on the Alamo, the city's four colonial missions, and their outlying ranchlands, its highest honor: status as UNESCO World Heritage sites. These landmarks were the first in Texas to be so distinguished and the twenty-third in the United States. The formal announcement, which capped a nine-year nomination process that the San Antonio Conservation Society and a slew of public and private partners developed, was touted as providing a much-needed economic boost to local tourism. One year later, the data seemed to confirm these initial aspirations. The National Park Service, which has managed the San Antonio Missions National Historical Park since 1993, indicated that between 2016 and 2017, more than 1.3 million visitors toured the park, dropping an estimated $110 million into the local economy and supporting as many as 1,300 jobs. That high dollar amount was a nice boost over the 2012 returns, in which tourists spent roughly $33 million in and around the park.[1]

There was more to the World Heritage designation than simply enhancing the city's time-honored pursuit of other people's disposable income, although this quest has been a critical part of San Antonio's strategy for growth since before the Civil War. More compelling is how this nomination sought to resurrect those who inhabited this region long before the Spanish arrived in 1718. The Payaya and other native nations, as hunter and foragers, had managed the San Antonio River valley for millennia. Their capacity to live within this bountiful landscape was circumscribed by the arrival of more powerful, horse-riding tribes from the north and the colonizing Spanish from the south. Caught between these

Mission Concepción exterior, from http://whc.unesco.org/en/documents/136237 [Accessed Mar. 1, 2018]. *Photo by Robert Howen, ©National Park Service.*

conflicting forces, the Payaya and other Indian peoples through choice, coercion, and persuasion entered the missions that Franciscan friars established in what we now know as San Antonio. The indigenous people's impress on these environments was significant. "The missions' physical remains comprise a range of architectural and archaeological structures including farmlands (*labores*), cattle grounds (*ranchos*), residences, churches, granaries, workshops, kilns, wells, perimeter walls and water distribution systems," the UN designation asserted. This array of structures and working landscapes demonstrates the "exceptionally inventive interchange that occurred between indigenous peoples, missionaries, and colonizers that contributed to a fundamental and permanent change in the cultures and values of all involved, but most dramatically in those

The striking face of Mission Concepción upon being "Restored by the Light." *Photo by Joan Vinson for the Rivard Report.*

of the Coahuiltecans and other indigenous hunter-gatherers who, in a matter of one generation, became successful settled agriculturists." Out of the dynamic exchange between those physical spaces and the people who lived within them emerged a "widespread sharing of knowledge and skills among their inhabitants, and the early adoption of a common language and religion resulted in a people and culture with an identity neither wholly indigenous nor wholly Spanish that has proven exceptionally persistent and pervasive." This complex integration was highlighted as well in the missions' architectural elements, combining "Catholic symbols with indigenous natural designs."[2]

One sign of this integrative culture came into focus in 2015 when projected on to the walls of Mission Concepción. That August, the city's Office of Historic Preservation collaborated with the National Park Service in recapitulating the mission's original, brilliantly colored façade, a dramatic realization.[3] Drawing on research that historic preservationists had been conducting on the site's frescos since the late 1920s, the nighttime light show, called "Restored by

Light," dazzled viewers with the intricate geometric patterns and bold palette that the Payaya and other laborers had painted on the eighteenth-century structures they called home. However ephemeral this projection, now an annual fixture in the city's cultural calendar, it offered a substantive illustration of the tight link between San Antonio then and now, and the pressing need to reclaim the community's fascinating and fraught heritage. This restorative impulse could not have been better timed: in 1718, as in 2018, the past was ever-present.

Notes

Chapter 1: Yanaguana

1. T. N. Campbell, *The Payaya Indians of Southern Texas* (San Antonio: Southern Texas Archaeological Association, 1975), 26.

2. Mattie Alice Austin Hatcher, *The Expedition of Don Domingo Terán De Los Ríos into Texas* (Austin: Texas Catholic Historical Society, 1932), 15.

3. Quoted in Frank W. Jennings, "Naming San Antonio 1691," *Journal of the Life and Culture of San Antonio* http://www.uiw.edu/sanantonio/jenningsnaming.html [Accessed Feb. 5, 2018].

4. Campbell, "The Payaya Indians of Southern Texas," 6–8; Karen E. Stothert, *The Archaeology and Early History of the Head of the San Antonio River* (San Antonio: Incarnate Word College, 1989), 44 (quotation).

5. Alvar Núñez Cabeza de Vaca, *The Narrative of Alvar Núñez Cabeza de Vaca*, ed. Frederick W. Hodge (New York: Charles Scribner's Sons, 1907), 66.

6. William Cronon, *Changes in the Land: Indians, Colonists, and the Ecology of New England* (New York: Hill and Wang, 1983), 41, makes the same case for how northern New England tribes managed their hunger.

7. James F. Patterson, "San Antonio: An Environmental Crossroads on the Texas Spring Line," in *On the Border: An Environmental History of San Antonio*, ed. Char Miller (San Antonio: Trinity University Press, 2005), 17–37.

8. Vinton Lee James, *Frontier and Pioneer Reflections of Early Days in San Antonio and West Texas* (San Antonio: Artes Grafica, 1938), 155.

9. Glenn J. Farris, "Depriving God and the King of the Means of Charity: Early Nineteenth-Century Missionary Views of Cattle Ranchers Near Mission La Purísma, California," in *Indigenous Landscapes and Spanish Missions: New Perspectives from Archaeology and Ethnohistory*, ed. Lee M. Panich and Tsim D. Schneider (Tucson: University of Arizona Press, 2014), 144. The most comprehensive analysis of indigenous fire-management practices is M. Kat Anderson, *Tending the Wild: Native American Knowledge and the Management of California's Natural Resources* (Berkeley: University of California Press, 2013).

10. Cabeza de Vaca, *The Narrative of Alvar Núñez Cabeza de Vaca*, 93.

11. Julianna Barr, "Beyond their Control: Spaniards in Texas," in *Choice, Persuasion, and Coercion: Social Control on Spain's North American Frontiers*, ed. Jesús F. de la Teja and Ross Frank (Albuquerque: University of New Mexico Press, 2005), 150.

12. Ibid., 152.

13. Andrés Resendéz, *The Other Slavery: The Uncovered Story of Indian Enslavement in America* (Boston: Houghton Mifflin Harcourt, 2016), 175.

14. Ibid. See also Pekka Hämäläinen, *The Comanche Empire* (New Haven: Yale University Press, 2009).

15. Barr, "Beyond their Control," 151 (first quotation); Jesus F. de la Teja, *Faces of Béxar: Early San Antonio and Texas* (College Station: Texas A&M University Press, 2016), 13 (second quotation).

16. Barr, "Beyond their Control," 151.

17. Fray Gabriel de Vergara's insights included in J. Villasana Haggard, "Spain's Indian Policy in Texas," *Southwestern Historical Quarterly* 46 (July 1942): 77.

18. Ibid., 76.

19. Tsim D. Schneider and Lee M. Panich, "Native Agency at the Margins of Empire: Indigenous Landscapes, Spanish Missions, and Contested Histories," in Panich and Schneider, *Indigenous Landscapes and Spanish Missions*, 55–56.

20. Campbell, "The Payaya Indians of Southern Texas," 2. Campbell indicates that in 1720 there were 26 Payaya at Mission San Antonio de Valero in San Antonio; a decade later there were 41 individuals; by 1780, only two. Ibid., 12.

Chapter 2: Urban Prospect

1. Fray Espinosa quoted Jennings, "Naming San Antonio 1691," <http://www.uiw.edu/sanantonio/jenningsnaming.html> [Accessed Feb. 2, 2018].

2. Fray San Buenaventura Olivares quoted in Jesús F. de la Teja, *San Antonio de Béxar: A Community on New Spain's Northern Frontier* (Albuquerque: University of New Mexico Press, 1995), 7. This and subsequent chapters of this book leans heavily on De la Teja's impeccable scholarship.

3. Ramón quoted in Karen E. Stothert, *The Archaeology and Early History of the Head of the San Antonio River* (San Antonio: Incarnate Word College, 1989), 54.

4. De la Teja, *San Antonio de Béxar*, 7.

5. Axel I. Mundigo and Dora P. Crouch, "The City Planning Ordinances of the Laws of the Indies Revisited. Part I: Their Philosophy and Implications," *Town Planning Review* 48 (July 1977): 248.

6. Ibid., 254.

7. Ibid., 248.

8. Ibid., 250.

9. Quotation from Gilberto M. Hinojosa, "The Religious-Indian Communities: The Goals of the Friars," in *Tejano Origins in Eighteenth-Century San Antonio*, ed. Gerald E. Poyo and Gilberto M. Hinojosa (Austin: University of Texas Press, 1991), 67.

10. Ibid., 70–71.

11. Governor Almazán quoted in Jesús F. de la Teja, "Forgotten Founders: The Military Settlers of Eighteenth-Century San Antonio," in *Faces of Béxar: Early San Antonio and Texas* (College Station: Texas A&M University Press, 2016), 56.

12. De la Teja, *San Antonio de Béxar*, 13.

13. Félix D. Almaráz Jr., "San Antonio›s Old Franciscan Missions: Material Decline and Secular Avarice in the Transition from Hispanic to Mexican Control," *The Americas* 44 (July 1987): 1n1.

14. Gerald E. Poyo, "The Canary Island Immigrants of San Antonio: From Ethnic Exclusivity to Community in Eighteenth-Century Béxar," in Poyo and Hinojosa, *Tejano Origins in Eighteenth-Century San Antonio*, 42

15. Ibid.
16. Ibid., 43.
17. Quoted in De la Teja, *San Antonio de Béxar*, 153.
18. Ibid., 11.
19. Fray Morfi quoted in ibid., 45.
20. Fray Morfi quoted in preface of Poyo and Hinijosa, *Tejano Origins in Eighteenth-Century San Antonio*, x.
21. Preface to Poyo and Hinijosa, eds., *Tejano Origins in Eighteenth-Century San Antonio*, x.
22. Poyo, "The Canary Island Immigrants of San Antonio," in ibid., 56.

Chapter 3: Revolutionary Space

1. "Introduction: Antonio Menchaca in Texas History," in *Recollections of a Tejano Life: Antonio Menchaca in Texas History,* ed. Timothy Matovina and Jesús F. de la Teja (Austin: University of Texas Press, 2013), 39.
2. Jesús F. de la Teja, "The Saltillo Fair and Its San Antonio Connections," in *Faces of Béxar: Early San Antonio and Texas* (College Station: Texas A&M University Press, 2016), 79–91; Elizabeth A. H. John, "Independent Indians and the San Antonio Community," in Gerald E. Poyo and Gilberto M. Hinojosa (eds.), *Tejano Origins in Eighteenth-Century San Antonio* (Austin: University of Texas Press, 1991), 126 (quotation).
3. Ibid., 127–128.
4. Jesús F. de la Teja, *San Antonio de Béxar: A Community on New Spain's Northern Frontier* (Albuquerque: University of New Mexico Press, 1995), 85.
5. Official Record (transcription), Jan. 19, 1793, <http://www.sonsofdewittcolony.org//adp/history/mission_period/valero/secular4.html> [Accessed Feb. 5, 2018].
6. Gerald E. Poyo, "Immigrants and Integration in Late Eighteenth-Century Bexar," in Poyo and Hinojosa, *Tejano Origins in Eighteenth-Century San Antonio*, 100. The Adaesanos reflect San Antonio's strikingly diverse population of that time, for their numbers included seven mestizos, five Indians, and only three Spanish. They were also highly conscious of their political rights, and repeatedly petitioned for arable land for more than a decade. In 1780, one of their leaders, Bernardo Cervantes, "a sixty-year-old Indian," even journeyed south to Chihuahua for an audience with the commander of interior provinces, to plead his people's case, a plea that reinforced the government's decision to secularize Mission San Antonio de Valero.
7. Félix D. Almaráz Jr., "San Antonio's Old Franciscan Missions: Material Decline and Secular Avarice in the Transition from Hispanic to Mexican Control," *The Americas* 44 (July 1987): 1–4.
8. Gilberto M. Hinojosa and Anne A. Fox, "Indians and Their Culture in San Fernando de Bexar," in Poyo and Hinojosa, *Tejano Origins in Eighteenth-Century San Antonio*, 119.
9. De la Teja, "Rebellion on the Frontier," in *Faces of Béxar*, 162–163.
10. Ibid., 163–164.
11. Laura Caldwell, "Casas Revolt," *The Handbook of Texas Online*, <https://tshaonline.org/handbook/online/articles/jcc02> [Accessed Feb. 5, 2018].
12. Robert H. Thonhoff, "Battle of Rosillo," *The Handbook of Texas Online*, <https://tshaonline.org/handbook/online/articles/qfr02> [Accessed Feb. 5, 2018]; De la Teja, "Rebellion on the Frontier," 169.
13. Mattie Austin Hatcher (trans.), "Joaquin de Arredondo's Report of the

Battle of the Medina, August 18, 1813," *The Quarterly of the Texas State Historical Association* 11 (January 1908): 220.

14. De la Teja, "Rebellion on the Frontier," 166–168; Hatcher (trans.), "Joaquin de Arredondo's Report of the Battle of the Medina," 220–236; Robert P. Marshall, "Archaeological Confirmation of the Site of the Battle of Medina: A Research Note," *Southwestern Historical Quarterly* 121 (July 2017): 56–66; Robert H. Thonhoff, "Battle of Medina," *The Handbook of Texas Online*, <https://tshaonline.org/handbook/online/articles/qfmo1> [Accessed Feb. 5, 2018].

15. De la Teja, "Rebellion on the Frontier," 168, 174–175.

16. Thomas F. McKinney to Stephen F. Austin, Sept. 9, 1829, in *Annual Report of the American Historical Association 1922: The Austin Papers*, Vol. 2, ed. Eugene Barker (Washington, D.C.: Government Printing Office, 1928), 256.

17. Bernice Strong, "José Francisco Ruiz," *The Handbook of Texas Online*, <https://www.tshaonline.org/handbook/online/articles/fru11> [Accessed Feb. 5, 2018].

18. Virginia H. Taylor (trans. and ed.), *The Letters of Antonio Martínez, Last Spanish Governor of Texas, 1817–1822* (Austin: Texas State Library, 1957), 241–243.

19. Ibid., 66; De la Teja, "The Colonization and Independence of Texas: A Tejano Perspective," in *Faces of Béxar*, 183.

20. De la Teja, "Colonization and Independence of Texas," 184.

21. Ibid., 185 (quotation).

22. This and following paragraphs follow De la Teja, "Colonization and Independence of Texas," 186–192.

23. Ibid., 190 (quotation).

24. Ibid., 191.

25. Juan Antonio Chávez interview, *San Antonio Express*, Apr. 19, 1914, reprinted in *The Alamo Remembered: Tejano Accounts and Perspectives*, ed. Timothy M. Matovina (Austin: University of Texas Press, 1995), 16.

26. Alwyn Barr, "Siege of Bexar," *The Handbook of Texas Online*, <https://tshaonline.org/handbook/online/articles/qebo1> [Accessed Feb. 5, 2018].

27. José María Rodríquez, "Memoirs of Early Texas," reprinted in Matovina, *Alamo Remembered*, 114; William Barret Travis to President of the Convention, Mar. 3, 1836, reprinted in Wallace O. Chariton, *Exploring the Alamo Legends* (Dallas: Republic of Texas Press, 2004), 201.

28. Eulalia Yorba interview, *San Antonio Express*, Apr. 12, 1896, reprinted in Matovina, *Alamo Remembered*, 54–57.

29. Stephen L. Hardin, *Texian Iliad: A Military History of the Texas Revolution* (Austin: University of Texas Press, 1994) offers the most detailed account of the battle and its military significance.

30. Ibid., 217.

31. All quotes in this paragraph come from David Montejano, *Anglos and Mexicans in the Making of Texas, 1836–1986* (Austin: University of Texas Press, 1987), 26–27.

32. David R. McDonald and Timothy M. Matovina (eds.), *Defending Mexican Valor in Texas: José Antonio Navarro's Historical Writings, 1853–1857* (Austin: State House Press, 1995), 20.

33. Juan N. Seguín, "A Foreigner in My Native Land," in Jesús F. de la Teja, *A Revolution Remembered: The Memoirs and Correspondence of Juan N. Seguín* (Austin: State House Press, 1991), 73–74.

Chapter 4: Forces of Americanization

1. Amy S. Greenberg, *A Wicked War: Polk, Clay, Lincoln, and the 1846 U.S. Invasion of Mexico* (New York: Vintage Books, 2012), 67–110.

2. Karl Jack Bauer, *The Mexican War, 1846–1848* (Lincoln: University of Nebraska Press, 1974), 146–147.

3. George W. Hughes, *Memoir Descriptive of the March of a Division of the United States Army, under the Command of Brigadier General John E. Wool, from San Antonio de Bexar, in Texas, to Saltillo, in Mexico* (Washington, D.C.: Government Printing Office, 1850), 9.

4. Samuel Chamberlain, *My Confessions: Recollections of a Rogue* (New York: Harper & Brothers, Publishers, 1956), 39 (quotation), 30–45.

5. Richard Eighme Ahlborn, *The San Antonio Missions: Edward Everett and the American Occupation, 1847* (Forth Worth: Amon Carter Museum, 1985), 7.

6. Hughes, *Memoir Descriptive of the March of a Division of the United States Army*, 10.

7. Char Miller, *Deep in the Heart of San Antonio: Land and Life in San Antonio* (San Antonio: Trinity University Press, 2004), 165–168; Lewis F. Fisher, "Preservation of San Antonio's Built Environment," in *On the Border: An Environmental History of San Antonio*, ed. Char Miller (San Antonio: Trinity University Press, 2005), 199–218.

8. Hughes, *Memoir Descriptive of the March of a Division of the United States Army*, 10.

9. Kevin R. Young, "Edward Everett," *The Handbook of Texas Online*, <https://tshaonline.org/handbook/online/articles/fev15> [Accessed Feb. 7, 2018]. Everett's design for the Alamo can be seen at <http://alamostudies.proboards.com/thread/695/everett-plan-depot-alamo-1848> [Accessed Feb. 7, 2018].

10. Everett was not entirely pleased with restoration work, especially the parapet: "I regret to see by a late engraving of this ruin, tasteless hands have evened off the rough walls, surmounting them with a r[i]diculous scroll, giving the building the appearance of a headboard or bedstead." Ahlborn, *San Antonio Mission*, 17.

11. Thomas T. Smith, *The U.S. Army and the Texas Frontier Economy, 1845–1900* (College Station: Texas A&M University Press, 1999), 29–33, 50–51.

12. Ibid.; for all the new wealth in San Antonio, by 1850 its merchant elite, mostly Irish immigrants, were not as well off as their peers in Galveston, Houston, or Austin. "San Antonio, the most populous of the four communities, was, by all standards, the poorest." See Kenneth W. Wheeler, *To Wear a City's Crown: The Beginnings of Urban Growth in Texas, 1836–1865* (Cambridge, Mass.: Harvard University Press, 1968), 110–111.

13. Quotation from Arnoldo De León, *They Called Them Greasers: Anglo Attitudes toward Mexicans in Texas, 1821–1900* (Austin: University of Texas Press, 1983), 82–83.

14. David R. McDonald and Timothy M. Matovina (eds.) *Defending Mexican Valor in Texas: José Antonio Navarro's Historical Writings, 1853–1857* (Austin: State House Press, 1995), 21.

15. Timothy M. Matovina, *Tejano Religion and Ethnicity: San Antonio, 1821–1860* (Austin: University of Texas Press, 1995), 89.

16. Ibid.

17. Frederick Law Olmsted, *A Journey Through Texas; Or, a Saddle-Trip on the Southwestern Frontier* (New York: Dix, Edwards & Company, 1857), 150–151.

18. To put it another way: "The San Antonio of Olmsted was quite different than the San Antonio of Juan Seguín only twelve years before"; see David Montejano, *Anglos and Mexicans in the Making of Texas, 1836–1986* (Austin: University of Texas Press, 1987), 29.

19. Andrew J. Torget, *Seeds of Empire: Cotton, Slavery, and the Transformation of the Texas Borderlands, 1800–1850* (Chapel Hill: University of North Carolina Press, 2016), 55.

20. *Journal of the Secession convention of Texas, 1861*, 88. Available online at <https://babel.hathitrust.org/cgi/pt?id=uc1.$b727365;view=1up;seq=99> [Accessed Feb. 7, 2018].

21. Raúl A. Ramos, *Beyond the Alamo: Forging Mexican Ethnicity in San Antonio, 1821–1861* (Chapel Hill: University of North Carolina Press, 2008), 229–230.

22. "The Revolution in Texas," *New York Times*, Mar. 2, 1861.

23. Roy Sylvan Dunn, "The KGC in Texas, 1860–1861," *Southwestern Historical Quarterly* 70 (April 1967), 543–573.

24. "The Revolution in Texas," *New York Times*, Mar. 2, 1861.

25. J. C. Houzeau, *La terreur blache au Texas et mon evasion* (Brussells: Ve Parent & fils, 1862).

26. Jean-Charles Houzeau, *My Passage at the New Orleans Tribune: A Memoir of the Civil War Era*, ed. David C. Rankin (Baton Rouge: Louisiana University Press, 1984), 10–12.

27. Quotation from Richard Parker and Emily Boyd, "Murder on the Nueces," *New York Times*, Aug. 11, 2012. See also Stanley S. McGowen, "Battle or Massacre? The Incident on the Nueces, August 10, 1862," *Southwestern Historical Quarterly* 104 (July 2000): 64–86.

28. Wheeler, *To Wear a City's Crown*, 154 (quotation).

29. Ibid., 154–158.

Chapter 5: Recovery and Development

1. Lewis F. Fisher, *American Venice: The Epic Story of San Antonio's River* (San Antonio: Maverick Publishing Company, 2015), 18–19.

2. Donald E. Everett, "San Antonio Welcomes the 'Sunset'—1877," *Southwestern Historical Quarterly* 65 (July 1961): 47–48; "San Antonio and Mexican Gulf Railroad," *The Handbook of Texas Online*, <https://The Handbook of Texas Online.org/handbook/online/articles/eqs08> [Accessed Feb. 8, 2018].

3. *Everett*, "San Antonio Welcomes the 'Sunset'—1877," 48 (quotation).

4. David R. Johnson, "Frugal and Sparing: Interests Groups, Politics, and City Building in San Antonio, 1870-1885," *Urban Texas: Politics and Development*, ed. Char Miller and Heywood T. Sanders (College Station: Texas A&M University Press, 1990), 33–57; Kenneth Mason, *African American Americans and Race Relations in San Antonio, Texas, 1867–1937* (New York: Garland Publishing, 1998), 86.

5. Mason, *African American Americans and Race Relations in San Antonio,* 83.

6. Johnson, "Frugal and Sparing," 33–57.

7. Ibid., 38.

8. *San Antonio Express*, Sept. 15, 1885, reprinted in Donald E. Everett, *San Antonio: The Flavor of its Past, 1845–1898* (San Antonio: Trinity University Press, 1975), 125.

9. Mason, *African American Americans and Race Relations in San Antonio*, 27, 31.

10. Ibid., 38–39; *San Antonio Express*, June 20, 1884, reprinted in Everett, *San*

Antonio, 79–80; Judith Berg Sobré, *San Antonio on Parade: Six Historic Festivals* (College Station: Texas A&M University Press, 2003).

11. Rev. Mark Henson quoted in Sobré, *San Antonio on Parade*, 60–61.

12. Char Miller, *Deep in the Heart of San Antonio: Land and Life in South Texas* (San Antonio: Trinity University Press, 2004), 119–120.

13. Ibid., 160–161; Everett, *San* Antonio, 6–7 (quotations).

14. Laura Hernández-Ehrisman, *Inventing the Fiesta City: Heritage and Carnival in San Antonio* (Albuquerque: University of New Mexico Press, 2008), 12–13.

15. Ibid, 15; Sobré, *San Antonio on Parade*, 163–194.

16. Everett, *San Antonio*, 7 (quotation).

17. Char Miller, "Where the Buffalo Roamed: Ranching, Agriculture and the Urban Marketplace," in *On the Border: An Environmental History of San Antonio*, ed. Char Miller (San Antonio: Trinity University Press, 2005), 56–70.

18. Ibid., 70–78.

19. Fredson Bower (ed.), *Stephen Crane: Tales, Sketches, and Reports* (Charlottesville: University of Virginia Press, 1973), 468–473.

20. Charles R. Porter, *Spanish Water, Anglo Water: Early Development in San Antonio* (College Station: Texas A&M University Press, 2009), 96–111.

21. Ibid, 113; Miller, *Deep in the Heart of San Antonio*, 89–91.

22. *San Antonio Express*, Jan. 10, 1893, reprinted in Everett, *San Antonio: The Flavor of its Past*, 59.

23. Bobbie Whitten Morgan, "George W. Brackenridge and His Control of the San Antonio Water Supply, 1869–1905" (master's thesis, Trinity University, 1961), 101 (quotation).

24. Heywood T. Sanders, "Empty Taps, Missing Pipes: Water Politics and Policy," in Miller, *On the Border*, 141–168.

25. Johnson, "Frugal and Sparing," *Urban Texas*, 55.

26. Marilyn McAdams Silbey, *George W. Brackenridge: Maverick Philanthropist* (Austin: University of Texas Press, 1973), 129, 150–156.

27. *San Antonio Express*, Jan. 7, 1883, reprinted in Everett, *San Antonio*, 71; see also Randall Lionel Waller, "The Callaghan Machine and San Antonio Politics, 1885–1912" (master's thesis, Texas Tech University, 1973).

28. John A. Booth and David R. Johnson, "Power and Progress in San Antonio Politics, 1836–1970," in *The Politics of San Antonio: Community, Progress, and Power*, ed. David R. Johnson, John A. Booth, and Richard J. Harris (Lincoln: University of Nebraska Press, 1983), 10.

29. *San Antonio Express*, Feb. 7, 1887, reprinted in Everett, *San Antonio*, 142–143.

Chapter 6: A New Day

1. President Theodore Roosevelt, text of speech in front of the Alamo, Apr. 7, 1905, <http://www.theodore-roosevelt.com/images/research/txtspeeches/132.txt> [Accessed Feb. 9, 2018].

2. Douglas Brinkley, *The Wilderness Warrior: Theodore Roosevelt and the Crusade for America* (New York: Harper, 2009), 313–314.

3. *San Antonio Express*, July 15, 1887, reprinted in Donald E. Everett, *San Antonio: The Flavor of its Past, 1845–1898* (San Antonio: Trinity University Press, 1976), 144.

4. *San Antonio Express*, May 20, 22, 29, 1898, reprinted in Everett, *San Antonio*, 75.

5. Brinkley, *Wilderness Warrior*, 314.

6. Brinkley, *Wilderness Warrior*, 317–319.

7. President Theodore Roosevelt, text of speech in front of the Alamo, Apr.7, 1905, <http://www.theodore-roosevelt.com/images/research/txtspeeches/132.txt> [Accessed Feb. 9, 2018].

8. Randall Lionel Waller, "The Callaghan Machine and San Antonio Politics, 1885–1912" (master's thesis, Texas Tech University, 1973), 19.

9. Christopher Long, "San Antonio Arsenal," *The Handbook of Texas Online*, <https://tshaonline.org/handbook/online/articles/qbs02> [Accessed Feb. 9, 2018].

10. John A. Booth and David R. Johnson, "Power and Progress in San Antonio Politics, 1836–1970," in *The Politics of San Antonio: Community Progress, and Power*, ed. David R. Johnson, John A. Booth, and Richard J. Harris (Lincoln: University of Nebraska Press, 1983), 11.

11. Erik Larsen, *Isaac's Storm: A Man, A Time, and the Deadliest Hurricane in History* (New York: Vintage, 2000).

12. Bradley Robert Rice, *Progressive Cities: The Commission Government Movement in America, 1901–1920* (Austin: University of Texas Press, 1977).

13. *San Antonio Daily Express*, Jan. 8, 1911.

14. Ibid.

15. Ibid.

16. Booth and Johnson, "Power and Progress in San Antonio Politics, 1836–1970," in Johnson, Booth, and Harris, *The Politics of San Antonio*, 13–16.

17. *San Antonio Express*, Oct. 3, 1913.

18. C. E. Ellsworth, *The Floods in Texas in September 1921* (Washington D.C.: Government Printing Office, 1923), 54; Lewis F. Fisher, *American Venice: The Epic Story of San Antonio's River* (San Antonio: Maverick Publishing Company, 2015), 36–41.

19. Fisher, *American Venice*, 42.

20. Ibid., 51–53. The Metcalf & Eddy report is summarized in Charles W. Sherman, "The Flood of September 1921 at San Antonio, Texas," *Proceedings of the American Society of Civil Engineers* 47 (November 1921): 801–802. Sherman had contributed to the 1920 Metcalf & Eddy report. See also Miller, *Deep in the Heart of San Antonio*, 61–66.

21. Fisher, *American Venice*, 54.

22. *San Antonio Light*, Dec. 17, 1887 reprinted in Everett, *San Antonio*, 144–145.

23. Char Miller, "Streetscape Environmentalism: Flood Control, Social Justice, and Political Power in San Antonio, 1921–1975," *Southwestern Historical Quarterly* 114 (October 2014): 159–177.

Chapter 7: All Quiet on the Southwestern Front

1. *San Antonio Express*, Nov. 11, 1919, 1–2.

2. Thomas A. Manning, *A History of Military Aviation in San Antonio* (Washington, D.C.: U.S. Department of Defense, 2000), 2.

3. "The Zimmermann Telegram," <https://www.archives.gov/education/lessons/zimmermann> [Accessed Feb. 14, 2018]; Ralph W. Steen, "World War I," *The Handbook of Texas Online*, <https://tshaonline.org/handbook/online/articles/qdw01> [Accessed Feb. 14, 2018].

4. "Wilson's Message to Congress," Apr. 2, 1917, <https://wwi.lib.byu.edu/index.php/Wilson%27s_War_Message_to_Congress> [Accessed Feb. 14, 2018].

5. Ibid.

6. Ibid.

7. "Formal U.S. Declaration of War with Germany, 6 April 1917," <https://en.wikisource.org/wiki/Formal_U.S._Declaration_of_War_with_Germany,_6_April_1917> [Accessed Feb. 14, 2018]

8. Art Leatherwood, "Leon Springs Military Reservation," *The Handbook of Texas Online*, <https://tshaonline.org/handbook/online/articles/qbl06> [Accessed Feb. 14, 2018].

9. Manning, *History of Military Aviation in San Antonio*, 9–10.

10. Ibid., 11.

11. George McCormac to Leona Perkins, Jan. 14, 1918, *The National World War I Museum and Memorial*, <http://theworldwar.pastperfectonline.com/archive/CA40205B-CE5B-4874-A34E-891866980917> [Accessed Feb. 14, 2018].

12. Ibid.

13. Nick Kotz, *The Harness Maker's Dream: Nathan Kallison and the Rise of South Texas* (Forth Worth: TCU Press, 2013), 89–92.

14. *San Antonio Light*, Nov. 12, 1918; *San Antonio Express*, Nov. 12, 1918.

15. Manning, *History of Military Aviation in San Antonio*, 15.

16. Woodrow Wilson, "Proclamation 1539—Thanksgiving Day, 1919," <http://www.presidency.ucsb.edu/ws/?pid=72445> [Accessed Feb. 14, 2018].

17. Richard A. García, *Rise of the Mexican American Middle Class: San Antonio, 1929–1941* (College Station: Texas A&M University Press, 1991), 255.

18. Ibid., 256–257.

19. Kenneth Mason, *African Americans and Race Relations in San Antonio, Texas, 1867–1937* (New York: Garland Publishing, 1998), 211–216.

Chapter 8: Turbulent Twenties

1. Char Miller and David R. Johnson, "The Rise of Urban Texas," in *Urban Texas: Politics and Development*, ed. Char Miller and Heywood T. Sanders (College Station: Texas A&M University Press, 1990), 17.

2. J. B. Guinn, "San Antonio—Flood City," *Survey*, October 8, 1921, 45–46.

3. *San Antonio Express*, Nov. 28, 1926.

4. Ibid., Nov. 21, 1926.

5. Char Miller and Heywood T. Sanders, "Olmos Park and the Creation of a Suburban Bastion" in Miller and Sanders, *Urban Texas,* 124–126.

6. *San Antonio Light*, Feb. 20, Feb. 21, 1907.

7.Char Miller and Heywood T. Sanders, "Parks, Politics, and Patronage," in Miller, *On the Border*, 83–90; John A. Booth and David R. Johnson, "Power and Progress in San Antonio Politics, 1836–1970," in *The Politics of San Antonio: Community Progress, and Power*, ed. David R. Johnson, John A. Booth, and Richard J. Harris (Lincoln: University of Nebraska Press, 1983), 3–27.

8. *San Antonio Express*, Aug. 3, 1919.

9. *San Antonio Express*, Nov. 23, 1923. For similar discussions concerning who would pay for much-needed flood control measures, see *San Antonio Express*, Dec. 3, 1923.

10. Park Acquisition data published by the City of San Antonio, Parks and Recreation Department, 1981 (in author's possession).

11. Bradley Rice, *Progressive Cities: The Commission Government Movement*

in America, 1901–1920 (Austin: University of Texas Press, 1977), 84–99; *San Antonio Light*, May 20, 1928, May 8, 1930. The vote margins in other black precincts were not as great as in precinct 54, but most provided comfortable enough margins that helped carry the election. Some of the Hispanic west side precincts were even more helpful: precinct 10, which the *San Antonio Light* indicated had been "a machine stronghold for years, and whose voters primarily are Mexicans, gave the bonds the greatest majority in any one precinct. The vote there stood 412 for and 60 against." See *San Antonio Light*, May 8, 1936. The machine clearly needed every Hispanic and black vote it could garner to win this particularly close election. Richard García, *Rise of the Mexican American Middle Class: San Antonio, 1929–1941* (College Station: Texas A&M University Press, 1991), 204–217.

12. *San Antonio Express*, May 11, 1927. Parks Commissioner Ray Lambert won by similar margins in precinct 54 and the other east side precincts.

13. Kenneth Mason, *African Americans and Race Relations in San Antonio, Texas, 1867–1937* (New York: Garland Publishing, Inc., 1998), 216–240, 265–278; Ralph Maitland, "San Antonio: The Shame of Texas," *Forum* 102 (August 1939): 53. Richard Henderson, *Maury Maverick: A Political Biography* (Austin: University of Texas Press, 1970), 178–179. Richard García in *Rise of the Mexican American Middle Class*, 204–217, suggests that west-side Tejanos were less politically cohesive and thus did not receive the kinds of political payback that the east side secured via Charles Bellinger.

Chapter 9: Deals Old and New

1. David M. Kennedy, *Freedom From Fear: The American People in Depression and War, 1929–1945* (New York: Oxford University Press, 1999), 48; Herbert Hoover, "Statement on Military Expenditures, July 23, 1929," at the *American Presidency Project*, http://www.presidency.ucsb.edu/ws/index.php?pid=21872. [Accessed April 1, 2018].

2. Julia Kirk Blackwelder, *Women of the Depression: Caste and Culture in San Antonio, 1929–1939* (2nd ed.; College Station: Texas A&M University Press, 1998), 18.

3. Lyle W. Dorset, *Franklin D. Roosevelt and the City Bosses* (Port Washington, N.Y.: Kennikat Press, 1977), 3–5, 112–116.

4. Robert B. Fairbanks, *The War on Slums in the Southwest: Public Housing and Slum Clearance in Texas, Arizona, and New Mexico, 1935–1965* (Philadelphia: Temple University Press, 2014), 36.

5. Quoted in ibid., 29.

6. Ralph Maitland, "San Antonio: The Shame of Texas," *Forum and Century* 102 (August 1939): 53.

7. Richard A. García, *Rise of the Mexican American Middle Class: San Antonio, 1929–1941* (College Station: Texas A&M University Press, 1991), 65.

8. Fairbanks, *War on Slums in the Southwest*, 51–54.

9. George Sessions Perry, "The Rumpled Angel of the Slums," *Saturday Evening Post*, August 21, 1948, 31–32, 43–44, 47. See also Char Miller, *Deep in the Heart of San Antonio: Land and Life in South Texas* (San Antonio: Trinity University Press, 2004), 119–122.

10. Karl Preuss, "Personality, Politics, and the Price of Justice: Ephraim Frisch, San Antonio's 'Radical Rabbi,'" *American Jewish History* 85 (September 1997): 275 (quotation); see also, 263–264; 269–277.

11. Ibid., 278–279.

12. Zaragosa Vargas, "Tejana Radical: Emma Tenayuca and the San Antonio Labor Movement during the Great Depression," *Pacific Historical Review* 66 (November 1997): 553–580; García, *Rise of the Mexican American Middle Class*, 64.

13. Archbishop Arthur Drossaerts quoted in John Weber, *From South Texas to the Nation: The Exploitation of Mexican Labor in the Twentieth Century* (Chapel Hill: University of North Carolina Press, 2015), 177.

14. Ibid., 178–180. This struggle also roiled Temple Beth El, for two of its congregants owned the Southern Pecan Company and some of the garment factories that were the focal points of labor unrest. Rabbi Frisch's sermons about his parishioners' unethical business practices were another source of controversy within the synagogue. See Preuss, "Personality, Politics, and the Price of Justice," 263–288; Nick Kotz, *The Harness Maker's Dream: Nathan Kallison and the Rise of South Texas* (Fort Worth: TCU Press, 2013), 133–134.

15. Judith Kaaz Doyle, "Maury Maverick and Racial Politics in San Antonio, Texas, 1938–1941," *Journal of Southern History* 53 (May 1987): 194–224; Richard B. Henderson, *Maury Maverick: A Political Biography* (Austin: University of Texas Press, 1970), 215.

16. Henderson, *Maury Maverick*, 215–217.

17. Lewis F. Fisher, "Preservation of San Antonio's Built Environment," in *On the Border: An Environmental History of San Antonio*, ed. Char Miller (San Antonio: Trinity University Press, 2005), 213.

18. Historic Sites Act of 1935, 49 Stat. 666; 16 U.S.C. § 461–467 (1935).

19. Henderson, *Maury Maverick*, 199–201.

20. Lewis F. Fisher, *Crown Jewel of Texas: The Story of San Antonio's River* (San Antonio: Maverick Publishing Company, 1997), 56.

21. Miller, *Deep in the Heart of San Antonio*, 169–171.

22. "Alamo Stadium: San Antonio, TX," <https://livingnewdeal.org/projects/alamo-stadium-san-antonio-tx-2/> [Accessed Feb. 19, 2018].

23. Ibid., 170–172.

24. Blackwelder, *Women of the Depression*, 183.

Chapter 10: The Economy of War

1. "A Snapshot of San Antonio During World War II: Blackout Simulation," <https://www.saconservation.org/VirtualExhibits/WWII/4CCF49D5-A2E6-4A1F-9DC8-220101429649.htm> [Accessed Feb. 19, 2018]; and "A Snapshot of San Antonio During World War II: Blackout Preparations," <https://www.saconservation.org/VirtualExhibits/WWII/F983420B-A854-41C2-93D6-296211239037.htm> [Accessed Feb. 19, 2018].

2. David R. Johnson, "San Antonio: The Vicissitudes of Boosterism," in *Sunbelt Cities: Politics and Growth since World War II*, ed. Richard M. Bernard and Bradley Rice (Austin: University of Texas Press, 1983), 235–236.

3. Thomas A. Manning, *A History of Military Aviation in San Antonio* (Washington, D.C.: US Department of Defense, 2000), 74–77.

4. Johnson, "San Antonio: The Vicissitudes of Boosterism," 236.

5. Manning, *History of Military Aviation in San Antonio*, 27–33, 88–89.

6. "A Snapshot of San Antonio During World War II: An International Airport," <https://www.saconservation.org/VirtualExhibits/WWII/8E6C986A-CB0E-4ECE-9940-181872698328.htm> [Accessed Feb. 19, 2018].

7. Melissa Gohlke, "Off-Limits and Out-of-Bounds: World War II and San

Antonio's Queer Community," *Top Shelf: Blog of the Special Collections of the UTSA Libraries*, Feb. 25, 2013, <https://utsalibrariestopshelf.wordpress.com/2013/02/25/off-limits-and-out-of-bounds-world-war-ii-and-san-antonios-queer-community/> [Accessed Feb. 19, 2018].

8. Julia Kirk Blackwelder, *Women of the Depression: Caste and Culture in San Antonio, 1929–1939* (College Station: Texas A&M University Press, 1984), 14, 24, 108.

9. Ibid.

10. Robert B. Fairbanks, *The War on Slums in the Southwest: Public Housing and Slum Clearance in Texas, Arizona, and New Mexico, 1935–65* (Philadelphia: Temple University Press, 2014), 47, 53–55.

11. Green Peyton, *San Antonio: City in the Sun* (New York: McGraw-Hill Book Company, Inc. 1946), 231.

12. R. Douglas Brackenridge, *Trinity University: A Tale of Three Cities* (San Antonio: Trinity University Press, 2004); Char Miller and Heywood T. Sanders, "Olmos Park and the Creation of a Suburban Bastion, 1927–1939," in *Urban Texas: Politics and Development*, ed. Char Miller and Heywood T. Sanders (College Station: Texas A&M University Press, 1990), 113–127; Donald E. Everett, *San Antonio's Monte Vista: Architecture and Society in a Gilded Age* (San Antonio: Maverick Publishing Company, 1999).

13. Robyn Ross, "A Sign of the Times in San Antonio," *Texas Observer*, June 5, 2012, <https://www.texasobserver.org/a-sign-of-the-times-in-san-antonio/> [Accessed Feb. 19, 2018].

14. Ibid.; Jack Morgan, "How World War II Changed History at San Antonio's Japanese Tea Garden," *Texas Public Radio*, Oct. 5, 2015, <http://tpr.org/post/how-world-war-ii-changed-history-san-antonios-japanese-tea-garden#stream/0> [Accessed Feb. 19, 2018].

15. Richard A. García, *The Rise of the Mexican American Middle Class: San Antonio, 1929–1941* (College Station: Texas University Press, 1991), 202–203, 298–299.

16. Ibid.

17. Nick Kotz, *The Harness Maker's Dream: Nathan Kallison and the Rise of South Texas* (Fort Worth: TCU Press, 2013), 188.

18. Howard E. Halvorsen, "Berlin Airlift: Faith in the Future," *Tinker* [Tinker AFB, Okla.] *Take-Off*, May 17, 2017.

19. Ann Markusen, Peter Hall, Scott Campbell, and Sabina Deitrick, *The Rise of the Gunbelt: The Military Remapping of Industrial America* (New York: Oxford University Press, 1991).

20. Manning, *History of Military Aviation in San Antonio*, 106–107.

21. James R. Compton, "Southwest Research Institute," *The Handbook of Texas Online*, <https://www.tshaonline.org/handbook/online/articles/sqs02> [Accessed Feb. 18, 2018].

22. Charles Gonzalez, "Henry B. Gonzalez—a life lived well, in service," *San Antonio Express-News*, May 1, 2016.

23. García, *Rise of the Mexican American Middle Class*, 316 (quotation); Jan Jarboe Russell, "Henry B. Gonzalez," *Texas Monthly*, January 2001,<http://www.texasmonthly.com/articles/henry-b-gonzalez/> [Accessed Feb. 19, 2018]; Kemper Diehl and Jan Jarboe, *Henry Cisneros: Portrait of a New American* (San Antonio: Corona Books, 1985).

24. Jeffrey Sullivan, "San Antonio Officially Becomes 'Military City USA,'" *The*

Rivar.d Report, July 19, 2017, <https://therivardreport.com/san-antonio-officially-becomes-military-city-usa/> [Accessed Feb. 19, 2018].

Chapter 11: Conflict, Consensus, and Change

1. Mary Beth Rogers, *Cold Anger: A Story of Faith and Power Politics* (Denton: University of North Texas Press, 1990), 125–126.

2. John A. Booth and David R. Johnson, "Power and Progress in San Antonio's Politics, 1836–1970, in *The Politics of San Antonio: Community, Progress and Power,* ed. David R. Johnson, John A. Booth, and Richard J. Harris (Lincoln: University of Nebraska Press, 1983), 18–25.

3. Ibid.

4. Randy Beamer, "Annexation: Land Grab or Prudent Plan?," Mar. 9, 2016, <http://news4sanantonio.com/news/san-antonios-voice/sa-annexation-land-grab-or-prudent-plan> [Accessed Feb. 20, 2018].

5. Enesto Cortés to Sterlin Holmesly (interview) in Sterlin Holmesly, *HemisFair '68 and the Transformation of San Antonio* (San Antonio: Maverick Pubishing Company, 2003), 119.

6. Char Miller, *Deep in the Heart of San Antonio: Land and Life in South Texas* (San Antonio: Trinity University Press, 2004), 157–171.

7. Rev. Claude W. Black Jr. to Holmesly (interview) in Holmesly, *HemisFair '68 and the Transformation of San Antonio*, 106.

8. Cortés to Holmesly in Holmesly, *HemisFair '68 and the Transformation of San Antonio*, 119.

9. Joseph D. Sekul, "Communities Organized for Public Services: Citizen Power and Public Policy in San Antonio," in Johnson, Booth, and Harris, *Politics of San Antonio,* 176.

10. Ibid., 177–178.

11. Char Miller, "Streetscape Environmentalism: Flood Control, Social Justice, and Political Power in San Antonio, 1921–1975," *Southwestern Historical Quarterly* 114 (October 2014): 159–177.

12. Ernesto Cortés Jr. to Lynnell J. Burkett, May 27, 1994 (interview transcript), pp. 4–5 (quotation), UTSA Libraries Digital Collections, <http://digital.utsa.edu/cdm/ref/collection/p15125coll4/id/275> [Accessed Feb. 20, 2018].

13. Gallegos quoted in Sekul, "Communities Organized for Public Services," in Johnson, Booth, and Harris, *Politics of San Antonio*, 175.

14. Miller, "Streetscape Environmentalism," 173–174.

15. John A. Booth, "Political Change in San Antonio, 1970–1982: Toward Decay or Democracy?," in Johnson, Booth, and Harris, *Politics of San Antonio*, 193.

16. Sidney Plotkin, "Democratic Change in the Urban Political Economy: San Antonio's Edwards Aquifer Controversy," in Johnson, Booth, and Harris, *The Politics of San Antonio*, 166–168.

17. Ibid., 160–161; Laura A. Wimberly, "Establishing 'Sole Source' Protection: The Edwards Aquifer and the Safe Drinking Water Act," in *On the Border: An Environmental History of San Antonio*, ed. Char Miller (San Antonio: Trinity University Press, 2005), 169–181.

18. Ibid. The State of Texas, with help from the Edwards Aquifer Authority, San Antonio Water System, the Trust for Public Land, and a variety of state and federal agencies, purchased the site in 1993; the public gained access to its trails and facili-

ties a decade later. See <https://tpwd.texas.gov/state-parks/government-canyon/park_history> [Accessed Feb. 20, 1018].

19. Plotkin, "Democratic Change in the Urban Political Economy," in Johnson, Booth, and Harris, *Politics of San Antonio*, 169; Charles Eaton Dunn and Hazel Dayton Gunn, *Reclaiming Capital: Democratic Initiatives and Community Development* (Ithaca, N.Y.: Cornell University Press, 1991), 128.

20. John M. Donahue and Jon Q. Sanders, "Sitting Down at the Table: Mediation and Resolution of Water Conflicts," in Miller, *On the Border*, 186.

21. Ibid.; Miller, *Deep in the Heart of San Antonio*, 71–96.

22. Nelson W. Wolff, who was involved in the behind-the-scenes negotiations as county judge for the Bexar County Commissioners, recounts his participation in the PGA Village imbroglio in his book *Transforming San Antonio: An Insider's Guide to the AT&T Center, Toyota, the PGA Village, and the River Walk Extension* (San Antonio: Trinity University Press, 2008), 50–86.

23. Wendy Holtcamp, "Saving Land, Saving Water," *Texas Parks & Wildlife*, June 2010, <https://tpwmagazine.com/archive/2010/jul/ed_1/index.phtml> [Accessed Feb. 20, 2018]. The San Antonio Edwards Aquifer Protection Program is detailed here: <https://www.sanantonio.gov/EdwardsAquifer/About> [Accessed Feb. 20, 2018].

24. Char Miller, "Faye Sinkin Changed City's Nature," *San Antonio Express-News*, Mar. 10, 2009.

Chapter 12: Future Shining?

1. J. Michael Kennedy and Karen Tumulty, "A 'Pink Elephant' Cisneros Cannot Forget," *Los Angeles Times*, Dec. 12, 1992.

2. Char Miller, *Deep in the Heart of San Antonio: Land and Life in South Texas* (San Antonio: Trinity University Press, 2004), 112–116.

3. Jan Jarboe Russell, "The Old Gray Mayor," *Texas Monthly*, December 2004, <http://www.texasmonthly.com/politics/the-old-gray-mayor/> [Accessed Feb. 21, 2018].

4. Nelson W. Wolff, *Mayor: An Inside View of San Antonio Politics, 1981–1995* (San Antonio: San Antonio Express-News, 1997); Wolff, *Transforming San Antonio* (San Antonio: Trinity University Press, 2008).

5. Jennifer R. Lloyd, "Trails renamed for former Mayor Peak," *San Antonio Express-News*, May 18, 2013.

6. "City South," DPZ, <http://www.dpz.com/Projects/0505> [Accessed Feb. 21, 2018]; Wolff, *Transforming San Antonio*, 190–191.

7. Robert Rivard, "A Land Bridge to Complete Hardberger Park," *Rivard Report*, Sept. 30, 2014, <https://therivardreport.com/land-bridge-unite-hardberger-park/> [Accessed Feb. 21, 2018]; Vianna Davila, "Hardberger Park Land Bridge close to fully funded," *San Antonio Express-News*, May 9, 2017.

8. Michael Barajas, "Your City is Going to Look Real Different, Real Soon," *San Antonio Current*, Aug. 24, 2016.

9. Julia Zorthian, "San Antonio's Mayor Blamed Poverty on 'People not Being in a Relationship with their Creator," *Time*, Apr. 27, 2017, <http://time.com/4757580/san-antonio-mayor-broken-people-religion-poverty/> [Accessed Feb. 21, 2018]. Taylor, who was appointed mayor to fill out the remainder of Julian Castro's term and won election in 2015, campaigned against her 2017 challenger, Ron Nirenberg, by attacking Castro, her predecessor. Her strategy backfired: Josh

Baugh and Vaanna Davila, "Nirenberg Defeats Taylor by Large Margin," *San Antonio Express-News*, June 11, 2017.

10. Thomas A. Manning, *A History of Military Aviation in San Antonio* (Washington, D.C.: US Department of Defense, 2000) details the local base realignments and closings up to 2000. For a general discussion of the national process of realignment, see Ann Markusen, Scott Campbell, Peter Hall, and Sabina Deitrick, *The Rise of the Gunbelt: The Military Remapping of Industrial America* (New York: Oxford University Press, 1991).

11. Vincent T. Davis, "Maria Berriozábal: The first Latina to serve on City Council," *San Antonio Express-News*, Sept. 9, 2015. Berriozábel's memoir, *Maria: Daughter of Immigrants* (San Antonio: Wings Press, 2012), recounts her lengthy career in public service and as an activist.

12. Wolff, *Transforming San Antonio*, 3–4.

13. Phil Hardberger, "City Council Members Deserve a Sensible Salary," *San Antonio Express-News*, May 4, 2015.

14. John Tedesco, "Losing Ground," *San Antonio Express-News*, Oct. 16, 2005.

15. Wolff, *Transforming San Antonio*, 5.

16. "San Antonio," United States Census Bureau, <https://www.census.gov/quickfacts/fact/table/sanantoniocitytexas/LND110210> [Accessed Apr. 1, 2016].

17. Amanda Peterka, "EPA Ozone Proposal Spells Doom for San Antonio's Compliance Streak," *E&E News,* April 6, 2015, <https://www.eenews.net/stories/1060016226> [Accessed Feb. 21, 2018]; Char Miller, "Oil and Water Don't Mix: What Fracking is Doing to South Texas," *KCET: Redefine*, Jan. 26, 2012, <https://www.kcet.org/redefine/oil-and-water-dont-mix-what-fracking-is-doing-to-south-texas> [Accessed Feb. 21, 2018].

18. Brian Chasnoff, "Study Foretold Abengoa's Woes," *San Antonio Express-News*, May 20, 2016; Chasnoff, "Water Talk Stays in 'Boardroom,'" *San Antonio Express-News*, Mar. 12, 2014.

19. Evan Hoopfer, "How will biggest Texas cities look in 2040?," *San Antonio Business Journal*, Oct. 20, 2016.

20. "Ron for Mayor Vision Booklet," < https://voteron.com/#body> [Accessed Feb. 21, 2018].

21. Joel Kotkin, "America's Next Great Metropolis is Taking Shape in Texas," *Forbes*, Oct. 13, 2016, <https://www.forbes.com/sites/joelkotkin/2016/10/13/the-next-great-american-metropolis-is-taking-shape-in-texas/#243c22df1e2f> [Accessed Feb. 21, 2018].

22. Melissa Stoelje, "A 'Tale of Two Countries,'" *San Antonio Express-News*, Jan. 6, 2017.

23. Kiah Collier, "Rich Schools Hopeful Houston ISD Could Topple Robin Hood Plan," *Texas Tribune*, Aug. 30, 2016, <https://www.texastribune.org/2016/08/30/rich-schools-hope-houston-topples-robin-hood-plan/> [Accessed Feb. 21, 2018].

24. Berriozábal, *Maria: Daughter of Immigrants*, 235.

25. O. Ricardo Pimentel, "A Tale of Two Downtowns and Hoping for a Happy Ending," *San Antonio Express-News*, Nov. 14, 2015.

26. John M. Donahue and John Q. Sanders, "Sitting Down at the Table: Mediation and Resolution of Water Conflicts," in *On the Border: An Environmental History of San Antonio*, ed. Char Miller (San Antonio: Trinity University Press, 2005), 194–195.

Restoration: An Afterword

1. "Visitor Spending Effects - Economic Contributions of National Park Visitor Spending," San Antonio Missions National Historical Park, 2016 data, <https://www.nps.gov/subjects/socialscience/vse.htm> [Accessed Feb. 21, 2018]. This data does not include the economic impact of the Alamo, which is not a part of the NPS site.

2. "San Antonio Missions," World Heritage List, <http://whc.unesco.org/en/list/1466> [Accessed Feb. 21, 2018].

3. "Seeing the Missions in a New (Old) Way," City of San Antonio Office of Historic Preservation, Aug. 30, 2016, <https://sapreservationstories.wordpress.com/2016/08/30/seeing-the-missions-in-a-new-old-way/> [Accessed Feb. 21, 2018].

Index